Palgrave Studies in Sound

Series Editor
Mark Grimshaw-Aagaard
Musik
Aalborg University
Aalborg, Denmark

Palgrave Studies in Sound is an interdisciplinary series devoted to the topic of sound with each volume framing and focusing on sound as it is conceptualized in a specific context or field. In its broad reach, Studies in Sound aims to illuminate not only the diversity and complexity of our understanding and experience of sound but also the myriad ways in which sound is conceptualized and utilized in diverse domains. The series is edited by Mark Grimshaw-Aagaard, The Obel Professor of Music at Aalborg University, and is curated by members of the university's Music and Sound Knowledge Group.

Editorial Board
Mark Grimshaw-Aagaard (series editor)
Martin Knakkergaard
Mads Walther-Hansen
Kristine Ringsager

Editorial Committee
Michael Bull
Barry Truax
Trevor Cox
Karen Collins

More information about this series at
http://www.palgrave.com/gp/series/15081

Seán Street

The Sound inside the Silence

Travels in the Sonic Imagination

Seán Street
Faculty of Media and Communication
Bournemouth University
Poole, UK

Palgrave Studies in Sound
ISBN 978-981-13-8448-6   ISBN 978-981-13-8449-3   (eBook)
https://doi.org/10.1007/978-981-13-8449-3

Cover image: oxygen

This Palgrave Macmillan imprint is published by the registered company Springer Nature Singapore Pte Ltd.
The registered company address is: 152 Beach Road, #21-01/04 Gateway East, Singapore 189721, Singapore

An attentive ear is the desire of a wise man
Ecclesiasticus 3:29 (KJV)

I hear it in the deep heart's core.
W. B. Yeats: 'The Lake Isle of Innisfree'

To Jo

Preface

In his work, *De Anima, (On the Soul)*, Aristotle begins his examination of the sense of hearing thus: '…Sound is in two ways, one in actuality, the other in potentiality' (Aristotle, 176). There is really no better way to explain the intent behind this book than that, and of the two sonic species identified by him, it is the second that in particular identifies the nature and essence of the journeys undertaken here. It is the third and last part of a trilogy that began with *Sound Poetics* and continued with *Sound at the Edge of Perception*. When the latter was published in 2018, a friend suggested that my books were growing quieter; if that has been in fact the case, and I suspect it is, then this writing is where silence takes over; at least, that is our starting point, but only insofar as it provides the blank canvas required for imaginative sound to create its pictures and transport us elsewhere.

Like its predecessors, this is a poetic exploration in at least two senses of the word; it will invoke the voices of writers and poets as guides to the sound world, and it will seek to explore its subject in the spirit of those voices, rather than as a writing of technical exposition. It is also a highly personal journey, because it is the only way I know of to ask some of the questions this book poses. In writing this, I have been trying to listen to myself listening, and what I think I have heard is something that can only be accounted for by the employment of an old fashioned and some would say overly romantic word: 'wonder'. To listen and to hear is one thing, to locate the whispers of response inside the imagination is another, but it would all be as nothing if it did not engender wonder at the music of everything, because it is that wonder that makes us go on listening to the world. Needless to say, as the journey has progressed, one question has provoked another, and then more; this is why I have found myself turning to the minds of so many others who have heard things and then tried to explain what it was that they heard. This too reflects the personal nature of the quest; on any long journey, it is reassuring to have friends around you, and I have found myself turning back to a number of familiar literary voices to provide moments of sonic illumination: Lucy Boston, Charles Dickens, Emily Dickinson, Robert Frost and Edward Thomas appear among others at various parts of the story, and for me, their perceptions help to clarify my own. Out of all these thoughts and ideas, my greatest wish in making this text is that something in it may speak with its own voice to those who hear poetry in sound, as well as sound in poetry, and who consider that listening is fundamental to understanding the world. Here, beginning with an open space, a field of consciousness as much as it is a piece of physical terrain, I hope we will people apparent silence (whether or not it actually exists—we shall debate that too) with sounds, music and imaginative possibilities that enables us to travel through time and space to the distant past, the infinite future and even to the afterlife, without stirring from where we are. Such is the wonder of the sonic imagination, because while sound is made up of events, it is our perception and psychic invention that gives it meaning and significance. Of course, it gets harder, this fishing for sound through active listening, because all the time, the world shouts louder and louder, and it is easier to shout over it than to wait and attend to the quiet voices beneath the cacophony, the whispers that may, in the end, be more important than anything else.

Sound recording, despite being with us since the nineteenth century, still only provides an auditory witness in technical terms to a fraction of history. On the other hand, human beings have been recording sound ever since they set foot on earth, through marks, paintings and words. It is in these that may be heard the voices of our ancestors and the world through which they moved, and there is much to learn from them. Likewise, the places in which we are rooted, begin and/or grow, are made of the sounds that fuel our trajectory into the world; the music of where we are is also the song of *who* we are, whether it be a pastoral hillside in dream England, or in downtown Memphis, Tennessee. Sound goes direct from heart to heart, which is why radio and its developing descendents continue to speak across boundaries and classes, opening imagination and painting colours that belong to everyone, while yet remaining intensely personal to each individual. We all carry our own sonic rainbow, and it is ours, because we made it out the raw materials of listening.

We take all this on the journey, just as we take the sound of the voices that made us, the dialects and grain drawn from region and town, and the people who, no longer physically heard, continue to travel alongside us in memory, speaking to us in the sound beneath the silence that is always there, always attentive and recording and playing back. We respond to memory, just as we hear with our eyes, emotionally; walk into an art gallery, and you join a continuing conversation that immediately includes you, triggering new ideas while at the same time acting as a mnemonic. Stand in front of a sculptural group, and listen, tune in; there are voices deep in the stone, striving to be heard by the sympathetic imagination. It is all simply a question of wavelengths; everything has a voice, although in the twittering world of media, it becomes harder to differentiate between truth and fake, between friendship and self-promoting oratory; the world can shout virtually as well as physically, and the mobile screens we carry in our pockets offer no respite. In this respect, hearing (in auditory terms or virtually) may be one thing, but listening is quite another; it is all about selection from the multifarious wavebands that bombard us.

We are in the end what we hear, or rather what we listen to and absorb; we are receivers as well as transmitters, and coming to the last full stop of this book, and to the conclusion of the journey that has led to it, there is still a sense that we have really only returned to the start, or rather, reached a new starting point, a new season of sonic awareness, a beginning for further imaginative travels. Waking up in Brussels one April morning during the 1930s to the sound of church bells, the French poet, dramatist and diplomat Paul Claudel heard in the carillons around him the coming of new possibilities emerging from a dark winter:

> The chimes have awakened and all sorts of voices in the air clamour for the right to speak. But, after a moment of this angelic stammering, silence falls solemnly. Something is going to happen… (Claudel, 64)

Liverpool, UK   Seán Street
Summer 2019

Acknowledgements

I would like to thank Piers Plowright and Cheryl Tipp for advice and insights, and also for permission to quote from correspondence. I am also grateful to Piers for allowing me to quote from his 2007 lecture, *Muzak, Music and Monologues—The Mind of Glenn Gould as Revealed in His Radio Documentaries*, given at the Canadian Museum of Civilisation's *Glenn Gould: The Sound of Genius* celebration in Gatineau–Ottawa. For their help, support and encouragement, it is a pleasure to thank Charlie Connelly, Richard Mabey, Robyn Ravlich, Professor Grażyna Stachyra of Maria Curie-Skłodowska University and Dr. Rachel J. Willie of Liverpool John Moores University. For her patience, support, and for reading and for commenting on the manuscript of this book, as always I offer thanks and love to my wife, Jo.

Contents

1

Silence Awaiting Sound: The Space of the Imagination

In Search of the Liminal

This is where it begins: with the tolling of a single bell. It is a favourite sonic image, starting in the violence of a strike and moving towards silence. Indeed, this is where it begins and *ends*, because the fade moves us from the physically heard sound to a point where we search our hearing for the disappearing tone in the air, and then in our imagination. Even some words for bells seem to mimic their action and purpose; in English, the word 'bell' begins with a plosive and mellows through its four letters until the continuant *L* sound seems to stretch softly at the tip of the tongue. In Spanish, 'La campana' begins with an attack and ends in '*aah*'; in French, 'cloche' is a '*clash*' that mellows to a whisper in the sibilant '*shh*'. It is little wonder that bells have connections with spiritual life, because they are messengers, metaphors and reminders. Sound vanishes, we find ourselves searching for its diminishing presence in the air, and even inside us, where its vibrations seem to persist, as some words do. In his 'Ode to a Nightingale', John Keats wrote:

© The Author(s) 2019
S. Street, *The Sound inside the Silence*, Palgrave Studies in Sound,
https://doi.org/10.1007/978-981-13-8449-3_1

Forlorn! The very word is like a bell
   To toll me back from thee to my sole self! (Keats, 209)

On the great tenor, bell in Winchester Cathedral is a Latin inscription: 'Horas Avolantes Numero, Mortuos Plango: Vivos as Preces Voco' ('I count the fleeting hours, I lament the dead, I call the living to prayer'. A previous book[1] ended with the sound of a bell, moving from sound to a place where sound—perhaps—continues in a mysterious other existence where we can detect no ending to its vibrations. This book continues the journey from where that story ended, seeking to explore not so much the sound of the bell itself, but the silence towards which it strives, and to search for the borderland between that sound and silence, the bridge where the ear ceases to hear but the mind is still listening. Winchester's bell is not alone in having an inscription: many bells are inscribed by their makers. It is almost as though we seek to rationalise in words the mystery of vanishing sound, and to hold its source within our physical world. There is a shape linked to function in all bells and that shape itself seems somehow symbolic: a sonorous space surrounded by material with which it interacts. To record the sound of a bell struck is to enable a process in which the initial dominant sound is gradually replaced by the peripheral ambience of the atmosphere and terrain in which it exists.

It is that liminality that will occupy us here, the apparent silence to which the very sound of the bell draws our attention, containing as it does the reality of the mind that perceives it. The Scottish writer Nan Shepherd expressed it in her book about the Cairngorms, *The Living Mountain*, 'such a silence is not a mere negation of sound. It is a new element, and if water is still sounding with a low far-off murmur, it is no more than the last edge of an element we are leaving, as the last edge of land hangs on the mariner's horizon…I am an image in a ball of glass. The world is suspended there, and I in it' (Shepherd, 75). In other words, we are concerned with the sound inside us, prompted as it may be by externals, the sonic response of the imagination to the world that surrounds it. The sound of a bell is something that travels through itself and across space and time, but it is rooted to the place in which it sings. Thus we begin our journey accompanied by that song in its various incarnations.

Silence might be considered to be an anechoic chamber—a space from which sound is locked out. Imagine a world like this. Consider a vacuum, an airless, soundless void. In a precursor to this work, *Sound Poetics*,[2] I wrote about negative silence: the silence of dissent, exclusion, social stigmatism, isolation and loneliness. Here, rather than beginning from the point of a failure in communication, let us think of a silence that can be a place of potential; renovating a house, we may wish to peel layers of old paper from the walls to return to a blank surface, upon which we place our true chosen colours. A bell may fade, but sound also emerges from silence. I am seeking to find an attentive silence upon which to place sonic colour, a place that is already present in the mind but that perhaps has been layered over by the circumstances of living in the material world. Or to use another, more positive analogy, perhaps we may see this silence as that of possibility, of a colour in itself. Let us imagine whiteness. William Hazlitt, in an essay written in the late eighteenth century called 'Why Distant Objects Please', wrote, 'the ear…is oftener courted by silence than noise; and the sounds that break that silence sink deeper and more durably into the mind' (Hazlitt, 125).

It is awareness that breaks the inner silence. Just as we breathe without thinking, so we hear without listening. When the brain attends to the ear's messages, the sounds that we had not noticed sinter into fragments, or rather into various notes and 'colours' of their own, like the divisions of a spectrum. The same is true of generalised noise, which in a way is its own silence; amorphous it may seem, but divided into its component parts, it may become readable. Proactive perception is *all* in this, because, like icebergs, so much of us lies hidden in the subconscious: memory and imagination, the ability to invoke active listening to these things. So perhaps it is stillness rather than silence out of which the sounds will come. On my desk is a Tibetan 'Singing Bowl'. As I run a striker around its rim, a pure tone gradually grows more intense. Where there was nothing, there is now sound, and it lasts as I circle the bowl with this piece of wood. It is music where there was nothing, working in the same way as a finger on the wet rim of a glass, or a glass harmonica. Let us then breathe atmosphere into space, ambience, because by so doing we make the world sonic and enable it to sing in its various voices. But even in the stillest of cells, there is the sound of the human mind, remembering, imagining. The human imagination is an active thing; it lies constantly present within us, but we possess the ability to exercise it and invoke it, as an athlete consciously tunes physicality. Listening likewise can be an either passive or 'switched on' at will. This will be our subject: the awareness of things based on the idea of how they are, or might be, at their core. My last book, *Sound at the Edge of Perception*,[3] ended with Emily Dickinson's poem, imagining her own deathbed:

> I heard a fly buzz – when I died –
> The Stillness in the Room
> Was like the Stillness in the Air –
> Between the Heaves of Storm…

Here, we will start from that self-same stillness, and ask how we are to fill it, even when all else falls silent. The answer must be through the imagination. Like the sound recordist, Gordon Hempton, I would advocate that this must be our entry point because 'good things come from a quiet place: study, prayer, music, transformation, worship, communion' (Hempton, 12). Yet locating this blank canvas is easier said than done in the modern world; just as pure darkness was a part of vision in the past, revealing stars and subtle gradations of shadow and light, so silence was more readable because our senses were more tuned to its presence. The writer, Lucy M. Boston, well known for her haunting books for children, each of which in their own way have the capacity to bridge time, was highly sensitive to sound and silence, and we shall call upon her perceptions more than once during these journeys. 'Pockets of silence still exist in blissful distant places', she wrote:

> The fascination is in the maintenance of what was a condition from the beginning of life, a natural beauty so taken for granted as to be unrecognised until it had gone. The present generation has no conception of silence. If it could be imagined it would be the silence of death, not of abounding life. Formerly it enfolded everything. We broke into it and it closed round us again. This gave great interest to sounds when they occurred, lost now since noise is the continuum. (Boston 1973, 10)

It is not just that today's is a noisy world, but that there is something within us that seeks out and contributes to the noise in it, as if to reassure ourselves that we are not alone, rather than being at ease with the act of waiting and listening for what matters. The Norwegian explorer Erling Kagge has suggested:

> The constant impulse to turn to something else – TV series, gadgets, games – grows out of a need with which we are born, rather than being a cause. This disquiet that we feel has been with us since the beginning; it is our natural state. The present hurts, wrote Pascal. And our response is to look ceaselessly for fresh purposes that draw our attention outwards, away from ourselves. (Kagge, 37)

We speak of *breaking the silence*, which implies both its almost physical nature and its fragility. A single sound placed in a great space of silence draws attention to that silence as well as to itself, as a pebble thrown into a lake draws attention to its sound and as well as to the water it enters by the ripples that animate the surface. Gradually, the ripples subside, but the eye remains fixed on the glassy reflections as they settle. The sound of the pebble, like the visible explosion of its entry point, vanishes, and the stillness, which is the water's natural state, slowly returns, and we find ourselves more aware of it than before. The lake, that was broken by the event of the pebble's attack, has mended itself. The sound that enters the silence breaks it, but the sound waves fade as they spread; the sound that audibly fades leaves the silence stronger and heightens consciousness of the minute constituent parts of that silence, which prove it to have been, in the end, no silence at all, just as the water on the lake was never actually still. The differences between looking and seeing, and hearing and listening are parallel and interconnected requirements in a true attempt to perceive the world around us. We shall talk in the next chapter about sound 'events' and our 'reading' of those events. How we respond to them may occupy us for a second and then be thought of no more, or turn a key that opens a magic moment of imaginative possibility, the start of a journey the limits of which are bounded only by the capacity of the mind for invention, which returns us to the sound of bells.

A bell has many moods and many meanings. It interacts with us in mysterious ways and lodges itself uniquely in the mind. It travels through air as it dies away, on its journey mixing with the same world that listens to its message. It is indeed a ubiquitous event both in town and in country, but when we hear them ringing in today's world, bells require an act of imagination to 'read' their sound. For a villager in the eighteenth or nineteenth centuries, the response would have been more instinctive as the sound came across their fields. Bells, as Alain Corbin reminds us, were *listened* to and evaluated according to a system of interpretation that is now largely lost to us.

> They bear witness to a different relation to the world and to the sacred as well as to a different way of being inscribed in time and space, and of experiencing time and space. The reading of the auditory environment would then constitute one of the procedures involved in the construction of identities, both of individuals and of communities. Bell ringing constituted a language and founded a system of communication that has gradually broken down. It gave rhythm to forgotten modes of relating between individuals and between the living and the dead. It made possible forms of expression, now lost to us, of rejoicing and conviviality. (Corbin, xix)

Corbin is writing of rural France, but the language of which he writes—that of the bells—knew no barriers other than the limits of human hearing. The poet Jean Ingelow, writing in the nineteenth century of an East Anglian flood that engulfed Boston in Lincolnshire in 1571, demonstrated powerfully through her poem 'The High Tide on the Coast of Lincolnshire' the capacity for a peel of bells to send out a code that would be quickly understood by the local community as a warning of impending disaster:

> 'Good ringers, pull your best,' quoth he.
> 'Play uppe, play uppe, O Boston bells!
> Ply all your changes, all your swells,
> Play uppe "The Brides of Enderby"'. (Ingelow, 146)

There was a message within the music that could broadcast certain information in a direct and quickly understandable way to communities within earshot. Even in the twentieth century, a British wartime populace was prepared to hear church bells, otherwise silenced for the duration, as a warning of impending invasion. We find other instances from literature to illustrate the various languages of bells; in Dorothy L. Sayers' detective novel *The Nine Tailors*, for example, it is a set of church bells that carry the plot, and in Leopold Lewis's melodrama, *The Bells*, made famous by Henry Irving in Edwardian Britain, it is the jingling of sleigh bells that haunts a murderer, and through his imagination, enters his conscience. Although much may have been lost, the nostalgic chord struck in us by the sound of bells across a meadow can still retain a power, particularly to the urban visitor, for whom their sound breaking into rural ambience, while the meaning may be lessened, is truly an 'event', and one which, on reflection, may signal an awareness of time, place and continuity: ('*They* would have heard that same sound in their day…'). The song of bells also continues to represent the communal identity of a place, for every church possesses its own timbre and music. Indeed, everywhere has its particular voice, rather like people, and under the shouting world, it is often still there to be heard, to the tuned ear.

I remember Henry Thoreau's description of sitting by Walden Pond and listening to sound at its various distances. One Sunday, he became particularly aware of church bells, their tones coming to him from various distances in surrounding communities—Lincoln, Acton, Bedford or Concord itself. He notices that, when the breeze is propitious, these sounds become part of the natural world; this is not intrusion, but a blending. A bell, after all, fades through the life of a single sounding towards silence, but the moment of inaudibility, where it melds with the ambient sounds around it, is subtle and elusive. Thoreau captures the idea perfectly:

At a sufficient distance over the woods this sound acquires a certain vibratory hum, as if the pine needles in the horizon were the strings of a harp which it swept. All sound heard at the greatest possible distance produces one and the same effect, a vibration of the universal lyre, just as the intervening atmosphere makes a distant ridge of earth interesting to our eyes by the azure tint it imparts to it. There came to me in this case a melody which the air had strained, and which had conversed with every leaf and needle of the wood, that portion of the sound which the elements had taken up and modulated and echoed from vale to vale. The echo is, to some extent, an original sound, and therein is the magic and charm of it. It is not merely a repetition of what was worth repeating in the bell, but partly the voice of the wood… (Thoreau 2016, 115)

Species of Silence

It is also true that all those things make imaginative sound that in turn fills the space that awaits them. 'Peace' and 'quiet' are words that together make a familiar phrase: 'How I long for some peace and quiet'. Yet it is much more than a negative empty space, because it is not a space at all, but a period of time into which thought pours with fewer interruptions. Hence a spawning ground for ideas. This moves us closer to the realms of the metaphysical, as expressed in the arts. The silence of which I would write is an expectation awaiting an event. It is an entity that descends when sound evaporates, and so is a form of sensual repose, yet at the same time it is darkness awaiting a light, containing the capacity to take us by surprise; it has its own inner power, potential. Silence can be accidental, or acquired. Our silence will be the silence John Cage's *4′ 33″* fills, radically different every time it is performed, changing like the air itself. Our stillness is not stasis, but the infinite silence of possibility. It is the receptacle that the rest of this book will seek to fill.

Seeking the most powerful example of an imaginative 'stretch' towards the idea of stillness, I turn back to twenty words in the Book of Revelation, itself probably the noisiest work in the entire Christian Bible, in which everything suddenly stops: 'And when he had opened the seventh seal, there was silence in heaven about the space of half an hour' (Chapter 8, verse 1). It is a moment of great mystery that has occupied the minds of mystics, historians, poets and theologians for centuries, not to mention sound artists (because it is so full of potential for the mind to take flight). It is all the more profound because of all the sounds going on in the text around it: screams, thunder, harps, trumpets, prophecies, praying and declaiming angels: a tumult of sonic colour, then suddenly this. The theological scholar Rachel Muers has asked the questions that must be in every mind that reads this section when she writes:

We are invited to pause and wonder, not only what happens next, but what this silence means. Why is there "silence in heaven" just at this point, and for just this length of time? Who is keeping silence, and for what end? What wider meaning of silence is brought into play here – wonder or terror, meaninglessness or fullness of meaning, suspense or completeness? Having begun to ask these questions, the reader might ponder them indefinitely, because the unexplained silence opens up so many possibilities for interpretation. (Muers, 1)

I am not a theologian, nor is this a journey into religious ideas; nonetheless, the questions posed by that one verse amongst so much else provide the perfect example of the silence of possibility that we possess as sentient beings. As always, the world plays its own music in duet with our imagination, and music of various kinds will be with us throughout this quest, so it is appropriate here to consider the words of two contrasting composers. 'Facing the silence of old trees', Toru Takemitsu felt that 'the quality of city life results in an abnormal swelling of the nerve endings' (Takemitsu, 3). Yet we carry our own transcendent space with us wherever we are and whatever we do; it is the power to invoke it that we require. As if answering his own question, Takemitsu quotes the great haiku master Matsuo Bashō:

As you look around
There is nothing
Which is not a flower. (ibid., 30)

In 1982, John Cage spoke to William Duckworth of his controversial work, *4 33″* as an act of listening rather than a performance: 'I listen to it constantly in my life experience. No day goes by without my making use of that piece in my life and work. I listen to it every day…I don't sit down to do it; I turn my attention toward it. I realise that it's going on continuously' (Cage in Gann, 186). Once we understand this, we are in a position to begin our journey. In order to conceive how the imagination fills so-called silence, we must first come to a realisation of the nature of silence itself, and its species, because it takes many and various forms. By actively attending to the possibilities of sonic space, we are already hearing things of which we may not be aware, linked to our own sentience. Listen to the room in which you find yourself and you become conscious of objects and forces that are always there, be they a fridge's cooling motor in the next room, a floorboard upstairs as someone walks on it, a branch brushing a window, or simply your own breathing, perhaps even—entering imagination, or sitting in an anechoic chamber—the pulse of your nervous system or the rush of blood through your veins. The apparently empty sound becomes occupied. Place yourself in another space, a place associated with history or momentous events, and listen again; the stillness around you evolves into its associative suggested stage whispers, full of imaginative messages created by your knowledge of where you are. Standing in the Amsterdam annex once lived in by Anne Frank and her family during the Nazi occupation of the Netherlands, I slowly became aware of what my mind was 'hearing', based on what I already knew of the happenings in those rooms. To this was added the ambience of the place itself, the murmur of other voices around me and then poignantly, the chime of a nearby church bell.

There are many forms of silence; it can be sacred, expectant, sorrowful, angry or oppressive, depending on how we ourselves inhabit it and interact with the other presences occupying it. Each species of silence is capable of producing an imaginative sonic response, both negative and positive. During the Spring of 1942, a little-known French satirist called Jean Bruller published *Le Silence de la Mer* under the pseudonym Vercors. It has become France's most enduring novel of the war and the German occupation. It has the intensity and compression of a prose poem and even in an English translation, conveys a powerful juxtaposition of stillness and suppressed violence. Its author explained his choice of title thus:

> I took a long time trying to find a title which would fit the hidden violence of this tale without sound or fury. Every day I lined up scores of them but found none to my liking. Then there came to mind a wild and poetic image which had often haunted me: beneath the deceptively calm surface of the sea, the ceaseless, cruel battles of the beasts of the deep. And I called my story *The Silence of the Sea*. ('Vercors', 25)

A German officer comes into the home of a Frenchman and his niece. From the start, their response to him is cloaked in silence, and it is a silence that contains many things at various stages of the story. When the officer speaks, his words fall into the silence. There is fear, anger, even repressed sexuality, but above all it is a tangible thing, that feels in large part created by the protagonists themselves: 'The silence was unbroken, it grew closer and closer like the morning mist; it was thick and motionless. The immobility of my niece, and for that matter my own, made it even heavier, turned it to lead' (ibid., 73). Later, as the soldier prepares to leave them, the silence returns, 'but this time how much more tense and thick!' It is here that 'Vercors' finds the metaphor for the whole story, the layers of silence that envelope them all:

> Underneath our silences of the past I had indeed felt the submarine life of hidden emotions, conflicting and contradictory desires and thoughts swarming away like the warring creatures of the sea under the calm surface of the water. But beneath this silence, alas! there was nothing but a terrible sense of oppression. At last his voice broke the silence. ('Vercors', 94)

Here, the employment of the idea of breaking a silence suggests in the very phrase the idea of a tangible thing that can be destroyed; this is a potent thought. In the first of his 'Glanmore Sonnets', the poet Seamus Heaney evokes a rural stillness that seems to emit from the land itself, 'a deep no sound' that reminds us of the silence 'like the morning mist...thick and motionless' of which Vercors writes. It presses on the sense, yet it is 'vulnerable to distant gargling tractors' (Heaney 1990, 109). In both cases, human intervention impinges on the silence. It becomes an event, something of which we will have more to discuss in the next chapter. We possess the capacity to create a silence that has stillness at its heart, and the very fact that these two words—silence and stillness—can occupy a sentence side by side is a demonstration of their kinship and differences. In 'Vercors' book, the silence encountered by the German, Werner von Ebrennac, is palpable, a silence of dignity and resistance, containing a psychic stillness that is almost measurable, something he can feel even as he approaches a room. Entering an empty house, moving through its rooms, we encounter various forms and degrees of silence.

By inhabiting a bedroom, for instance, we invest it with our own presence. An imaginative partnership between ourselves and a physical environment shapes a stillness we feel ourselves when we enter it, because prolonged privacy generates the essence of silent stillness. In his novel, *Bruges-la-Morte* Georges Rodenbach writes of a living silence that seems to pour out of houses and move along the canals, a tangible subtle presence in itself that seems abused, corrupted, desecrated by the invasion of any sound, which seems to render itself by its very presence as something with a grossness and inappropriate crudeness: in short, sound seems insolent and sacrilegious. Many writers have used the potentials of silence, as a receptacle.

Thus, place would seem to actively generate silence, while at the same time, our emotional engagement with an environment creates an imaginative dialogue, even though it is a wordless one. We have the capacity to impose our own inner silence on a space or a situation, but the world can get its own back on a chosen personal silence, by invading it with its own identity. It is for this reason that we must foster the art of listening at a deeper and more active level than, for the most part, our daily lives allow. Many have debated as to whether or not actual silence exists at all; stillness on the other hand is perhaps somewhat easier to identify, and a short list might begin with the following:

– The stillness of a place from which we come.
– The stillness we make and/or create out of mood, an active form of expression.
– The stillness we recognise when we listen actively.
– The stillness that is always inside us.

This last is important, because it is the stillness that finds kinship in all other forms, that recognises itself mirrored in the circumstances of living. The intimacy of a place echoes our true presence in it. This is the stillness of anticipation, of intense attention, of waiting for an answer. In November or December 1798, William Wordsworth wrote a poem which begins 'There was a Boy'. It recalls his childhood, and his habit of cupping his hands to his mouth and blowing, imitating the sound of the owls in the countryside around him, seeking to stimulate an answer from them:

> And they would shout
> Across the watery vale, and shout again,

Responsive to his call, – with quivering peals,
And loud halloos, and screams, and echoes loud
Redoubled and redoubled; concourse wild
Of jocund din! (Wordsworth, 183)

Wordsworth is in many ways *the* poet of sound, a tuned and active listener to the natural world, and to his own inner reactions to it. This poem, which opens the collection, *Poems of the Imagination* is typical in this sense; but it is in what he writes next that he exercises his great ability to bridge the physical and the imaginative worlds, rather like our bell:

And when there came a pause
Of silence such as baffled his best skill:
Then sometimes, in that silence, while he hung
Listening, a gentle shock of mild surprise
Has carried far into his heart the voice
Of mountain-torrents; or the visible scene
Would enter unawares into his mind
With all its solemn imagery, its rocks,
Its woods, and that uncertain heaven received
Into the bosom of the steady lake. (ibid.)

There is the stillness of water again, 'the steady lake', a visual metaphor for sonic stillness, waiting to be broken. The poem has moved the boy from the maker to the receiver; once the effort of exercising himself as a mimic has been exhausted and silence asserts itself, as Seamus Heaney has written, 'there occurs something more wonderful than owl-calls. As he stands open like an eye or an ear, he becomes imprinted with all the melodies and hieroglyphs of the world; the workings of the active universe, to use another phrase from *The Prelude*,[4] are echoed far inside him' (Heaney 2002, 227).

In the paintings of Edward Hopper too, silence becomes visible, and in the poetry of Paul Valery, it is like fog, a noise that 'blankets everything, this sand of silence…Now nothing. This nothing is huge in the ears' (Valery, quoted Corbin 2016, 4). Already, through these examples, we are entering the imaginative expression of others, and in so doing we are in a sense pre-empting and predicting aspects of this journey with which we will engage later. The extraordinary thing about silence is that it is intensely personal, and this makes for huge creative possibilities. Every silence is our own, awaiting our imagination's capacity for filling it with our own rainbow of meaning. We carry the place we occupy around with us, wherever we go, and it is subject to change according to mood and circumstance. As Corbin says, 'there are houses that breathe silence, where it seems to permeate the walls…Inside houses, various types of silence impregnate rooms, halls, bedrooms and studies' (ibid., 5). Nonetheless, we only hear the silence and give it meaning when we enter the space ourselves. Likewise, we may find ourselves in the centre of huge sound, be it a club with thumping base frequencies, a chattering restaurant or an echoing shopping mall, and yet feel a deep silence inside us, be it a void of emptiness, loneliness or a cultivated stillness of inner calm. The sound within us can be what we make it, or it can be imposed on our consciousness in spite of ourselves. To be able to creatively manage and filter our inner sound can be a liberating feeling; on the other hand, to find ourselves out of control and invaded by unwanted noise may be a deeply troubling, disturbing and dehumanising situation. It is the difference between invitation and invasion: the sudden harsh ringing of a telephone for example. We may feel impelled to answer it, or to ignore its command to respond, as if it were a rude person who without warning has burst into the room and demanded that we listen to them, irrespective of whatever act or conversation they were interrupting. To control silence, and the forces that impinge upon it, is a key part of understanding ourselves.

Tuning In

With the creation of radio came an external medium to break physical silence, with the capacity to fill the mind with imaginative sound and possibilities. The exploration of this power will be a key part of this study; sound informing thought comes out of a kind of sonic darkness and inhabits it, as a blank canvas may be imagined to 'hold' a drawing within itself while at the same time making its existence possible. The famous 1938 CBS broadcast adaptation of H. G. Wells's novel, *The War of the Worlds*, is a much-cited example of this. Just as the book created pictures in the mind in a pre-electronic era, so the use of the immediacy of a new form of journalism informed an idea that turned science fiction into a possible truth, fuelled by external forces such as international tensions and paranoia. In the days when wireless valves took time to warm up, the silence between switching on and hearing was invested with expectation and anticipation which contributed to the willing suspension of disbelief, just as the dimming of theatre lights before curtain-up prepares us for entry to strange and magical worlds.

Thus, into this bowl of stillness are placed sounds which have the quality of events, however small. Placed where they are by circumstance, they provide us with the character and context of our particular situation in the world. This is a dimension that is temporal, because a sound progresses through time (like the message of a bell). We shall make the imaginative leap into the past to gain a sense of how the sound of a place has changed, and what our ancestors heard that was different to our own experience—and what has been shared across the years. It is a music that is shaped by its location, and because it is in our nature to seek expression, we do so, often in the sounds we form out of our partnership with a place. Thus, a key element of the exploration this book will attempt is the music born out of the duet between human beings and their homeland. Sound, however, requires no passport to travel, neither does the imagination. We sense whispers coming from shadows, and what we do not know, we imagine; if we have no proof of what lies beyond the stars or the grave, we can make aural symptoms of these things within us. Be it ideas of the supernatural, or the possibilities of science fiction, imaginative sound has found some of its most eloquent expression in the storytelling that takes us into these worlds.

We make dream sounds from our own lives and pasts. Memory is and always has been the kind of imagination, and the reception and transmission of who we are are linked to our family history, and the poignant recollection of voices that we may remember, or have preserved through technology; playback can comfort, console or unsettle us, but whatever its effect, it forms the link to our personal history, and we hear our loved ones through the archive of the mind. A picture can evoke the memory of a voice; even the apparently solid form of a statue, a sculpture or a framed painting or photograph can create an imaginative sonic response in us. We shall spend time considering the effects of public art, the stones and metal forms with which we have peopled our landscapes, and ask ourselves what it is about them—their presence and as often as not, their silence—that speaks to us. Such stillness as the human-made additions to natural terrain is accentuated by the babble we make in our everyday lives; a particular concern here will be in the 'noise' of social media; there has always been opinion, argument and debate, but the phenomenon of imposed and sometimes created personas, and at times their ability to bully and shout through our computer and telephone screens, is something from which any study of imaginative sound cannot shy away. A text message contains the voice of its maker and provides a mnemonic that triggers various associations that can trouble our days and disturb our sleep. Considering this, we will move towards the conclusion of our journey and situate it where we have begun, within the mind itself.

The positive and negative forces of which we are shaped take many forms. We may 'hear' colour and 'see' sound, both physically and psychically. Because our senses are linked, one triggers others. Interactivity is at the heart of this study, and through time and activity, we have developed skills and art forms that blend voice and vision. Expression through the medium of dance, for instance, could be considered to be a visual art, yet of course it is linked to music and above all rhythm. It is sound made visible, interpreting aural patterns with shapes in space. A film of a dancer, viewed silently, may often convey the tempo of the sound to which the dance is created; memory may even supply the actual music if it is familiar. Some forms, such as tap, clog or Spanish, have the capacity to be dances for the blind, since a key ingredient lies in its sound. Dance is movement to music, but it could be said that tap and its kindred forms are dance *as* music. The dancer Paul Draper put it well when he said, 'What the eye sees is sharpened by what the ear hears, and the ear hears more clearly that which sight enhances' (Draper in Seibert, p. 4). With the eyes closed and the sound turned down, there is a void; slowly introduce the sound of percussive feet, and the image enters the mind, sonic light fills the consciousness and a picture evolves.

This book, then, is about the art of listening, and we have begun with apparent silence and stillness because it is within this creative space that what we hear is digested. We hear with every part of our bodies, through vibration, suggestion, curiosity. Our whole being is engaged in responses to sound, from the soles of our feet to our upper body, the abdomen, solar plexus, chest and head. Our ears are the tip of a highly tuned resonating chamber which is both receiver and musical instrument, and listening actively is the beginning of learning to play it. In a brutalised and brutalising world that increasingly shouts to itself, we must regain the ability to attend to stillness and focus on what it contains. The sonic imagination must centre itself upon what is speaking to it, as well as delving and searching for intense access to what lies beneath and behind it, and somehow rid itself of the instinct to analyse too deeply. When we over-think, we destroy the music of what is happening to us. The stillness may be awoken by something infinitely small, but when did loudness begin to assert itself as a key factor in sonic existence? The beginnings are usually almost untraceably subtle, but we need to be aware of them; a symphony begins with a single note, and it all *matters* profoundly. Seamus Heaney, writing of the poetry of John Clare, marvelled at his instinctive response to the minutiae of experience in the natural world. Clare did not think twice and allow the intellect to overmaster first responses; he acted as a conduit and what delivered itself to the page, stayed there, with every immediate nuance and apparently unimportant detail. Yet as Heaney wrote, it was this that connected him to 'the here-and-nowness, or there-and-thenness of what happened. I am reminded of a remark made once by an Irish diplomat with regard to the wording of a certain document. "This," he said, "is a minor point of major importance"' (Heaney 2002, 278).

There are many parallels between the acts of seeing and hearing, and that relationship will be crucial to this journey. Almost everything that relates to the tuning of the visual sense may be applied to the art of listening. This is why the visual arts play such an important part in enabling the sonic imagination. When Nan Shepherd wrote that the 'changing of focus of the eye, moving the eye itself when looking at things that do not move, deepens one's sense of outer reality', she was invoking the conscious act of looking and seeing, so that 'then static things may be caught in the very act of becoming' (ibid.). While we do not have quite the same control over our ears as our eyes, we may learn the skill of allowing audio consciousness to search downwards through layers of sound, as Shepherd's eye moved across the landscape, absorbing and learning from *re-seeing* what she thought she had already seen. Casting her eye across the high terrain of her beloved

Cairngorms through clear unpolluted air, she identified mountains and hills into the farthest distance. 'At midsummer', she wrote, 'I have had to be persuaded I was not seeing further even than that. I could have sworn I saw a shape, distinct and blue, very clear and small, further off than any hill the chart recorded' (ibid., 2). A trick of the light? And if so, can the ear be so fooled? Samuel Purchas, writing in his *Microcosmus, or the historie of man* of 1619, placed the ear and the eye as partners in terms of witness to times past and present:

> The eye seeth onely things present; the eare, by tradition of fathers to their children, receives the wisedome of our forefathers, and of those that are furthest remote both in time and place from us: and by speech and writing (a visible speech) the learning of the world is continued from the first man to the last; and this short age of man is by the eare, in manner, made immortall. (Purchas, quoted by Woolf in Smith, 120)

In some ways, this statement by Purchas could be said to encapsulate many of the themes of this book. Through thought, memory and imagination, we become time machines. It could also be added that the silence upon which imaginative sound might be said to lie is in itself illusory. If thinking is 'to be', then what is it that articulates thought within us but internal language, through the medium of speech. I think in English, which pre-supposes I am thinking in words, as speech. As the philosopher, Maurice Merleau-Ponty put it: 'To Kant's celebrated question, we can reply that it is indeed part of the experience of thinking, in the sense that we present our thought to ourselves through internal or external speech' (Merleau-Ponty, 206). As soon as we catch ourselves thinking, we have broken the silence.

Notes

1. Street, Seán, *The Poetry of Radio: The Colour of Sound.* Abingdon: Routledge, 2014.
2. Street, Seán, *Sound Poetics: Interaction and Personal Identity.* Cham: Palgrave, 2017.

3. Street, Seán, *Sound at the Edge of Perception: The Aural Minutiae of Sand and Other Worldly Murmurings.* Singapore: Palgrave, 2019.
4. *The Prelude: Or Growth of a Poet's Mind*, a long autobiographical poem completed by Wordsworth in 1805.

2

Fencing the Horizon: Sound as Imaginative Event

The Experience of Listening

Having identified our imaginative space, with all its potential for receptivity, we must now begin to people it. Sound and vibration—the sonic and the sense of touch—place us in the world, and imagination and memory are affected by the dynamics of the sound events around us. Each sound as it impinges upon our consciousness identifies itself as an event within the mind; a specific sound, be it a blackbird, someone coughing or the memory of a steam train hissing, is a sonic event placed on the canvas of the ambience of a place. John Berger expressed this idea in a short essay called 'Field' in which he contemplates a familiar rural space, listens to its silence and gradually, consciously heightens his own awareness of the small sounds that inhabit it:

> Into this concentrated tiny hub of dense silent noise, came the cackle of a hen from a nearby back garden…the noise of the hen, which I could not see, was an event…in a field which until then had been awaiting a first event in order to become itself realisable. I knew that in that field I could listen to all sounds, all music…The experience which I am attempting to describe…is very precise and is immediately recognisable. But it exists at a

© The Author(s) 2019
S. Street, *The Sound inside the Silence*, Palgrave Studies in Sound,
https://doi.org/10.1007/978-981-13-8449-3_2

level of perception which is probably proverbial -— hence, very much, the difficulty of writing about it. (Berger 1980, 200)

In that last sentence, Berger identifies an issue for himself as a writer, with which I can sympathise in the creation of this current text. We are exploring areas of great subtlety and nuance here; but Berger can help us:

> Remember what it was like to be sung to sleep…The repeated lines of words and music are like paths. These paths are circular and the rings they make are linked together like those of a chain. You walk along these paths and are led by them in circles which lead from one to the other, further and further away. The field upon which you walk and upon which the chain is laid is the song. (ibid., 199)

Harder still is to reach an understanding of the *experience* of listening and hearing those sounds, within the context of the time and the communities that made them and heard them. As we progress, we shall be required to stretch our imaginations into lost time; even some of the sounds of ten years ago are receding into memory, such is the pace of technological change. In order to proceed further back to a time before recording technologies, we must allow the texts of those who lived then to provide the witness and the clues, just as their ears transmitted an awareness of these events to their minds and thence to the page. Necessity breeds invention, and the desire to communicate experience through words has produced some of the greatest forms of expression human beings have achieved. Poetry is about—and made of—sound, but it is rooted in the need to show us a place, an idea or an emotion; language can be transcendent, but it begins with a concept, an impression, borne out of an experience. We may see (or rather hear) Berger's field as a metaphor for the space within the mind that awaits sensations of particular interest to us being in this context, the reception of sound. Falling rain is a visual phenomenon…until it strikes a surface, when it becomes auditory. Heard on a roof rather than seen, it provokes a unique set of responses within us. The moment of its sonic interaction with tile, road or tree is thus an event. Some may find that to play a recording of rain—the composite of multiple tiny incidents of impact—is an aid to sleep, relaxation or meditation. On the other hand, for some it may evoke a sense of fear, dread or confusion. The 'radio' of rain can thus be more subtly evocative in various ways than can the 'television' of the eyes in their attempts at seeing it. Likewise, the first impression of a field may be an event in visual consciousness. Yet the sound of a human cry, or a dog's bark, a plough or an unknown bird call emanating from within it, may change our response to it, superimposing a new sonic event upon the visual, as the airwaves of the place are moved and disturbed, rippling across intervening space and into our awareness. Minute sound interventions emanating from any source may move consciousness on from its contemplation of apparent silence, or conversely draw attention to it as a surrounding presence. In 1851, Henry David Thoreau, walking on a late afternoon in July, noted that 'I hear the peewee in the woods, and the cuckoo reminds me of some silence among the birds I had not noticed' (Thoreau 2009, 61). For Thoreau, the associations between sounds and a place sometimes became so intertwined that they merged into a single experience; a few days later, in early August, he comments that 'as my eye rested on the blossom of the meadow-sweet in a hedge, I heard the note of an autumnal cricket, and was penetrated with the sense of autumn. Was it sound? Or was it form? Or was it scent? Or was if flavour?' (ibid., 65). The memory of the event will merge these impressions in his mind even more as time moves him away from the moment. Human presence adds to the unpremeditated music of the natural world; Thoreau was fascinated by the coming of the telegraph, and of the strange sounds emitting from the wires, complemented by the wind as it brushed them into accidentals, like an Aeolian Harp. During the 1850s, this new phenomenon in the landscape was to him a sympathetic partner of nature, a man-made sound that did not intrude, but rather merged with the surrounding song of the landscape. It is September 1851:

> Yesterday and today the stronger winds of autumn have begun to blow, and the telegraph harp has sounded loudly. I heard it especially in the Deep Cut this afternoon, the tone varying with the tension of different parts of the wire. The sound proceeds from near the posts, where the vibration is apparently more rapid. I put my ear to one of the posts, and it seemed to me as if every pore of the wood was filled with music, laboured with the strain, – as if every fibre was affected and being seasoned or timed,

rearranged according to a new and harmonious law. Every swell and change of inflection of tone pervaded and seemed to proceed from the wood, as if its very substance was transmuted. What a recipe for preserving wood, perchance, – to keep it from rotting, – to fill its pores with music! (ibid., 81)

We are capable of forming a response to where we are, placed in the present but conscious that there is a history and a future to everything and everywhere, and the sound and feel of things change in tandem with their visual aspect. What may have been a field located once on farmland becomes surrounded by suburbs: traffic rumbles around it, radios or household appliances and machinery from nearby buildings may murmur or shout, sing or chant unwontedly. The sound of communal and territorial events takes ownership of the continuing event of the place itself, notes on its musical stave, all feeding into memory to be interpreted by imagination as our personal means of identifying the character of the immediate environment as it shapes us.

The Scream

The poet Elizabeth Bishop was born in Worcester, Massachusetts, in 1911. After her father died when she was just eight months old, her mother, Gertrude, became mentally ill. Elizabeth Bishop as she developed, recognised and profoundly understood the significance of sound events, and incorporated such moments throughout her poetry and prose, keys to unlock place and memory. In an early essay, 'Time's Andromedas', written during the 1930s while she was a student at Vassar College, New York, she offers a hint of a preoccupation with sound overtaking other senses. In it, she writes of trying to study as concentration lapses:

My own thoughts, conflicting with those of the book, were making such a wordy racket that I heard and saw nothing – until the page before my eyes blushed pink. I was startled, then realised that there must be a sunset at my back, and waited a minute trying guess the colour of it from the colour of the little reflection. As I waited I heard a multitude of small sounds, and knew simultaneously that I had been hearing them all along – sounds high in the air, of a faintly rhythmic irregularity, yet resembling the retreat of innumerable small waves, lake-waves, rustling on sand. (Bishop in Millier, 45)

The sound of migrating birds becomes an event that transforms the moment, but only when its presence, which had been there all along, is consciously acknowledged. From her earliest student days, Bishop's great poetic influence was Gerard Manley Hopkins, and this is significant because Hopkins's poetry is the epitome of sound through words. It was his writing above all others that taught her that a sound event can be a phrase or even a word, placed in juxtaposition with others to create a telling audio moment, just as two musical notes adjacent to one another can have the ability to stir the emotions spontaneously. Likewise, she demonstrates in this early passage how tuned she was as a sounding-board for the sonic signals around her. It was an instinct that seems to have been with her from her earliest years. Bishop had been brought up through her childhood by her grandparents in Great Village, Nova Scotia, initially with her mother, until Gertrude was committed to an asylum in 1916, when Elizabeth was five years old. She never saw her again. This potent time is described vividly in Bishop's masterpiece of a short story, *In the Village*. Here, we meet a five-year-old girl, who witnesses her mother's final breakdown in the context of the continuity of the small Nova Scotian village the boundaries of which formed her personal horizons. The story is framed by sound events: first, Gertrude's scream of fear as a dressmaker fits her first colourful dress after five years of mourning her husband. It is a sound that lingers in the air for the child, and, as a memory, throughout the life of the grown woman:

A scream, the echo of a scream, hangs over that Nova Scotian village. No one hears it; it hangs there forever, a slight stain in those pure blue skies…The scream hangs there…unheard, in memory – in the past, in the present, and those years between. It was not even loud to begin with, perhaps. It just came there to live, forever – not loud, just alive forever. Its pitch would be the pitch of my village. Flick the lightning rod on top of the church steeple with your fingernail and you will hear it. (Bishop, 251)

Shortly after this event, Elizabeth's mother is taken away. The recollected shock of the moment opens the door to a journey through the village, as the young child walks the family cow to pasture, and we gain a clear and poignant picture of the place, and the lonely and bewildered little girl within it: 'An immense, sibilant, glistening loneliness suddenly faces me, and the cows are moving off to the shade of the fir trees, their bells chiming softly, individually' (ibid., 265). Written in 1953, some 37 years after the event, for Bishop there is another, more comforting sound, towards which her story moves. A favourite haunt for the child in the story is the blacksmith's shop by a stream close to the family home, which Bishop remembered as being at the end of the garden, and thus a constant sonic presence. The village blacksmith was Mayhew (Mate) T. Fisher, who, in the story, becomes 'Nate'. The smithy was a haven of stability for her, and in the story, she begs him to make her a ring, which was to become, to the mature poet a metaphor for the transformative power of art. As the scream opens the story, it is the familiar domesticity of the blacksmith's shop that ends it, and in which it ultimately resolves itself, offering—perhaps or perhaps not—some kind of closure and resolution:

> *Clang.*
> Nate is shaping a horseshoe.
> Oh, beautiful pure sound!
> It turns everything else to silence.
> But still, once in a while, the river gives an unexpected gurgle. *"Slp,"*
> it says, out of glassy-ridged brown knots sliding along the surface.
> *Clang.*
> And everything except the river holds its breath (ibid., 274).

One sound event seems to almost obscure the other, redeeming memory. Surely now the scream has evaporated into the earth, or is fading far out to sea beyond Halifax harbour? As time passes, sounds, both feared and loved, become lost; we may feel relief or regret, but the continuity of the natural world replaces them in the senses, perhaps in the end, the only reality?

> It is the elements speaking: earth, air, fire, water.
> All those other things – clothes, crumbling postcards, broken china;
> things damaged and lost, sickened or destroyed; even the frail almost-

lost scream – are they too frail for us to hear their voices long, too
mortal?
Nate!
Oh, beautiful sound, strike again! (ibid.)

The sound event fades almost literally like a bell, and it is poignant
because it is an audio metaphor for ourself, and for life and experience.
The linear nature of sound—its arrival, be it sudden or emergent, its presence as it passes us and its gradual disappearance—creates sonic events that
alter a space in the mind and create, if not something new, at least a happening in which, like a photograph of an intimate environment, a familiar
routine or a traumatic incident, a moment that becomes printed on the
imagination and memory for ever afterwards. The circle of a small set of
horizons—in Bishop's case epitomised by her own village and the uni-
versality of childhood experience—frames and holds the sonic experience
powerfully, so much so that place and sound become one in recollection.
Once read, we carry her sound with us, committing them to our memory through the medium of words, in the story. Now, were we to visit
the locations she describes more than a century on from the event, we
would 'hear' Nate's long-gone blacksmith's shop, and even perhaps in a
lull between traffic on the busy trunk road, imagine a faint scream still
lingering in the air.

The World in Microcosm

In the circumscribed physical world of a child, every moment and every
experience are of immense importance. In terms of place, the environment
of a small community is analogous with this, because the ear that is familiar
to the sonic habits and occurrences of hamlets, villages, enclosed spaces
and even small towns is tuned to detect anomalies and novelties of sound
that might pass by the casual hearer. It is in this that the difference between
hearing and listening comes into play in a significant way. Jean-Claude
Nancy has suggested that 'if "to hear" is to understand the sense,…to listen
is to be straining toward a possible meaning, and consequently one that is
not immediately accessible' (Nancy, 6). Like Elizabeth Bishop, the English

poet Edward Thomas was extraordinarily sensitive to the sounds under sounds, and the minutiae of micro-sounds within apparent silence. In his book, *The Icknield Way*, he captures a magical moment in the village of Sparsholt in Hampshire, England. Part of his skill is in the way he sets the scene; he tells us that it is just after the children have gone inside the village school after morning play, a time when rural stillness intensifies, the road 'quieter than the church on that hot, bright morning'. It is then that he becomes conscious of distant music. The initial spell, his growing awareness of the sound, is formed by the context in which it is heard:

> As I walked under the garden trees I came slowly within hearing of a melody played so lightly, or so far lost among winds or leaves, that I could hardly distinguish it. It was an hour when nearly everyone is at work. A poor, ragged girl was walking in front of me in awkward haste. But she stopped at the same time as I did, to listen. The music was not everything. The shadow and the filtered light, the silence of the music half submerged, the busy hour so steeped in tranquillity, helped the player to express perfect carelessness and freedom from the conditions of life – summer, wealth, luxury, happiness, youth, gaiety, innocence, and benevolence. (Thomas 1913, 255).

This quality of deep listening, of the sound emerging from silence, is present in much of Thomas's prose writing, even in the early work. In *Oxford*, published in 1903, we find him at one point in a college garden, listening to the trees:

> Now and then they talk a little, and when one talks, the others follow; but as a rule the wryneck or the jackdaw talks instead; and with them it seems to be near the end of the day, nothing remaining save *benedictus benedicat*. In the angriest gale and in the scarcely grass-moving air of twilight the cypresses nod almost without sound. (Thomas [1932], 213)

We shall return to the sound of Edward Thomas's world in the next chapter. We shall also examine a powerful example of lost sound from the pen of Gilbert White, the eighteenth-century Hampshire vicar who immortalised his small village of Selborne through his minute and ground-breaking observations of the natural history he saw around him in his garden and within the surrounding hills and valleys. To grow where one is planted, so to speak, and live through seasons in the same place over a lifetime, is to gain understandings that through the local transmute to the universal. In the tradition of White is the contemporary American writer, Annie Dillard, whose book of observations of her surroundings at Tinker Creek in a valley in the Blue Ridge of Virginia is at once a close-wrought examination of her immediate environment around her home, and an interrogation of herself and her responses to it. She describes the house as being clamped to the side of the creek itself like an anchor. 'It's a good place to live; there's a lot to think about'. Within the narrow compass of an intensely focused place, Dillard finds there to be 'the mystery of the continuous creation and all that providence implies: the uncertainty of vision, the horror of the fixed, the dissolution of the present, the intricacy of beauty, the pressure of fecundity, the elusiveness of the free, and the flawed nature of perfection' (Dillard, 4/5). It is a place where ear and eye can recalibrate from the long, distant view of the mountains on the horizon, to the close-up dramas that play themselves out before her senses every day. Like Elizabeth Bishop, Dillard understands that a visual cue can provoke an internal sonic response, audible only to her as emotion, but still sound nonetheless. To return to Nancy, 'we listen to someone who is giving a speech we want to understand, or else we listen to what can arise from silence and provide a signal or sign, or else we listen to what is called "music"' (Nancy, 6). As we listen to the spoken word, we are programmed to make sense of it, and the words are the code by which we do so. Nancy would suggest that music is different because in this case 'it is from sound itself that sense is offered to auscultation. In one case, sound has a propensity to disappear; in the other case, sense has a propensity to become sound' (ibid., 6–7). Our senses feed our intellect and our emotions, and the sounds—all sounds—that surround us become a form of music within this definition. If we accept that the interpretation of the world we bring to ourselves through thought possesses some kind of silent sound within us, we are better placed to understand Annie Dillard's response to Tinker Creek when she writes, 'I walk out; I see something, some event that would otherwise have been utterly missed and lost; or something sees me, some enormous power brushes me with its clean wing, and I resound like a beaten bell' (Dillard, 14).

There are many examples of writers and poets whose compass was geographically narrow, while the inference they gained from it was universal. The Irish poet, Patrick Kavanagh, from County Monaghan, expressed it well when he wrote:

> To know fully even one field or one land is a lifetime's experience. In the world of poetic experience it is depth that counts, not width. A gap in a hedge, a smooth rock surfacing a narrow lane, a view of a woody meadow, the stream at the junction of four small fields – these are as much as a man can fully experience. (Kavanagh, in MacFarlane, 63).

The potency of ingrained familiarity is distilled in intimate space. Sound becomes part of the décor with which we surround ourselves so that any change registers in our consciousness as a new event, requiring itself to be identified and interrogated. John Berger's field is both a physical place and a field of consciousness, present all around us but awaiting recognition. The most intimate physical space of all is the home, the rooms we ourselves occupy on a daily basis. The acoustic of a room reflects its contents, and we become so aware of it as we move within a house that we cease to notice it, and may fail to sense it as a form of horizon imbued with human presences, until a sound event that does not belong to the routine of its specific text demands imaginative exploration. Just as Nan Shepherd's world was the Cairngorms, a terrain of which she became so familiar that 'I listened to the waterfall until I no longer heard it' (Shepherd, 8), likewise we must consciously reawaken the consciousness of the particular character in a familiar intimate space. Move around your living area and *hear* the rooms, the sonic perspective and presence they offer the ear in relation to what infiltrates from the wider horizon. Even their silences contain individual personalities; this place is like a well, into which a dropped sound is a happening that awakens walls. Just as we become aware of the minutiae of sound in a meadow, a wood or a garden, so by attending to stillness in a room, we hear it as layers that contain our own self, a microcosmic world that we create in our personal likeness. The intimate becomes universal through the atmosphere of which we have become a part as daily existence, and it peoples the confines within walls with a kind of immensity. Even a prison cell is a world when a single person's presence characterises it.

Emily Dickinson wrote all her work within the narrow physical horizon of her Amherst home, yet the immensity of her vision as expressed in hundreds of concise poems amounts to a distillation of human consciousness. It always comes down to observation, awareness, listening and looking, reducing experience to an internal space that is the most significant horizon of all, because by knowing it, we place ourselves within it, at its centre. Gaston Bachelard wrote that 'immensity is within ourselves. It is attached to a sort of expansion of being that life curbs and caution arrests, but which starts again when we are alone. As soon as we become motionless, we are elsewhere' (Bachelard, 184). Yet that 'elsewhere' is always present, and focusing on the day-to-day of a living environment enables us to hear it again. Sitting in the stillness of a known house, the creak of a floorboard, the fall of a letter on the doormat, the purring of a washing machine, the gentle 'bump' of a heating system as it ignites register as part of our essential soundtrack. An alien sound, say a bird's wing on the window, or a step in the hall when, by rights, there should be no one there—these awaken the audio consciousness in an almost primeval way; what is this that is moving in our personal space? Until it is identified, explained and remedied, it is a source of the uncanny and potential fear. Bachelard's 'elsewhere' is where we are, but changed by the invasion of new experience. Even the silence of a room is a sonic event in itself.

Weather

Right at the start of his novel, *Bleak House*, Charles Dickens paints a picture of London fog that is as vivid as if it were on canvas, perhaps more so. He does so to lure us through it, firstly, into mystery, and subsequently to a story that then opens with the metaphorical fog of seemingly endless legal wrangling in the impenetrable matter of Jarndyce and Jarndyce, and the bogged-down impasse of a tedious case in the House of Chancery. Does not, after all, the man who knows more of it than anyone else even carry the name of Mr Tangle? No wonder the book opens in an impenetrable London 'pea-souper':

Fog everywhere. Fog up the river, where it flows among green aits and mead-ows; fog down the river, where it rolls defiled among the tiers of shipping, and the waterside pollutions of a great (and dirty) city. Fog on the Essex marches, fog on the Kentish heights. Fog creeping into the cabooses of collier-brigs; fog lying out on the yards, and hovering in the rigging of great ships; fog drooping on the gunwales of barges and small boats. Fog in the eyes and throats of ancient Greenwich pensioners, wheezing by the firesides of the wards; fog in the stem and bowl of the afternoon pipe of the wrathful skipper, down in his close cabin; fog cruelly pinching the toes and fingers of his shivering little 'prentice boy on deck. Chance people on the bridges peeping over the parapets into a nether sky of fog, with fog all around them, as if they were up in a balloon, and hanging there in misty clouds. (Dickens 1903, 1–2)

I like to consider the possibility that Dylan Thomas had been reading *Bleak House* before embarking on the opening speech in *Under Milk Wood*. Certainly, both works begin in silence, before gradually expanding into action. Be that as it may, Dickens has given us a prose poem of the weather, in which our imagination journeys through his London with him, encountering a meteorological phenomenon, too common in his day, lurking and loitering in corners and crevices like a malign beast. Coming with the fog comes an inhuman stillness, because fog is nature's anechoic chamber.

*   *   *

Here and now, it is suddenly raining, beating against the glass, drumming on the roof. The squall has come from nowhere, imposing the event of itself on the mind. Weather is the most ubiquitous of conscious happenings, and a thunderclap may perhaps be the most familiar of its dramatic sonic events for most of us, something we can imbue with metaphor and human meaning that hints at a deep past of superstition and awe. The wind beats in the trees, and we experience it both as the sound of a storm and as a conscious emotional response. Wind is air made visible through the physical objects it affects, and in the sound, it makes in partnership with those objects. The bleakness of the storm is epitomised by the sound of the wind as it funnels through guttering or a downpipe. Weather can assault us, colour our mood, can reflect our joy or sorrow, while actually enhancing it. Unlike fog, rain can provide a shape to our terrain through its interaction with surfaces, as described earlier. John Hull was forced to embrace the world anew after becoming blind, and the character of rain and its percussion interacting with his surroundings shaped space for him, as he expressed in a memorable passage in which he considered the specific sound events as the rainwater struck the various surfaces around him:

> I hear the rain pattering on the roof above me, dripping down the walls to my left and right, splashing from the drainpipe at ground level on my left, while further over to the left there is a lighter patch as the rain falls almost inaudibly upon a large leafy shrub. On the right, it is drumming, with a deeper, steadier sound upon the lawn. I can even make out the contours of the lawn, which rises to the right in a little hill. The sound of the rain is different and shapes out the curvature for me. (Hull, 26)

Hull, having lost his sight, learned to interpret place in terms of the species of interaction between rain and the impact of it of surfaces. We may all do this, but it requires some reflection to analyse it in terms of the experience of environment. Listen, and you start to hear perspective. The rain is falling all around, so that the splashing sound on the path beats in a very different way to the absorbent sound on soil, or the rustle and rush on water. Sometimes it is ambiguous. Is it raining, or is that sound the wind in the trees? For a sighted person, opening a window provides a picture of the world that enables interpretation; what is the weather like? Is it busy out there? Is that someone coming up the path? For Hull, the event of rain 'presents the fullness of an entire situation all at once, not merely remembered, not in anticipation, but actually and now. The rain gives a sense of perspective and of the actual relationships of one part of the world to another' (ibid., 27).

Writers often consider storm, wrack and rain. These are, after all, events of drama which draw our attention, make us feel the frail things we are and seek for meaning and reason in chaos, when, in fact, there may be none. A calm, sunlit day is an idyllic film of a dream. Lightning and a thunderbolt is a photograph. While we may find drama in elemental sound events, we might strive to justify order and meaning in anthropomorphic terms. In a previous book, *Sound Poetics*,[1] I quoted from Tim Robinson's extraordinary work, *Connemara*, in which he describes the effect of weather on minute objects and man-made structures on the Irish coastline. It is an amazing description of elemental forces, and I return to it often. As he endures the moment, he considers the morning after, when, perhaps, 'a tumult of air will be battering the windows, all its wavelengths, from the vast heft of gusts over the hill that half shelters us, to the spasm of the garden shrubs and the fluting of a dry leaf caught between two stones, merging into one toneless bulk noise… Such vast, complex sounds are produced by fluid generalities impacting on intricate concrete particulars' (Robinson, 2). The images he creates as a reflection of the event are amazingly powerful and graphic, and the head fills with huge imaginary buffeting sounds. It is noise, yet even here we may find meaning. Jacques Attali presents us with a paradox: '…The very absence of meaning in pure noise or in the meaningless repetition of a message, by unravelling auditory sensations, frees the listener's imagination. The absence of meaning is in this case the presence of all meanings, absolute ambiguity, a construction outside meaning. The presence of noise makes sense, makes meaning' (Attali, 33). Of course there are metaphors; Tim Robinson is writing about one place and in it, the world. He is concerned with the history of it and of its people. In this context, he sees the destruction caused by the storm as analogous with the obliteration of truth of the sounds of the past, absorbing and sometimes drowning history's real voice:

> History has rhythms, tunes and even harmonies; but the sound of the past is an antagonistic multiplicity. Sometimes, rarely, a scrap of a voice can be caught from the universal damage, but it may only be an artefact of the imagination, a confection of rumours. (ibid.)

Be that as it may, we shall strive after those scraps of witness in our next chapter. There is, too, another metaphor which we might extract from the teeth of the storm; all the sounds that Robinson describes at the beginning of *Connemara* are, although blended to an amorphous mass by the contingency of weather, yet individual objects with their own identity, and each of their sounds on their own, however minute, is an event with its own audio existence. By being drawn into the shouting of this baleful choir, the constituent parts are beyond our hearing, and yet we do hear them, merged within the rest. Sound has become noise, and this noise is preventing us from listening to the individual voices, the component sonic events of which it is comprised. Attali may suggest to us that the noise still contains the meaning, but this only makes the effort of seeking it harder, while always worthwhile, and there are increasing forces in modern life that come between ourselves and the ability to identify it. If finding sound were archaeology, there remains a lot of digging to do.

When the sound breaks silence's ground, however, there is no denying that something has happened. Perhaps the greatest description of a sonic event I know was written by an American ex-patriot Scotsman, John Muir. Muir was born in Dunbar in 1838, but he emigrated with his father to the USA in 1849. He was to develop a love affair with America's great wild spaces that would last a lifetime, and today, he is still revered as a father of the National Parks and a pioneer of the conservation movement. He was many things in his lifetime: an inventor, a mountaineer, explorer, botanist, geologist and encompassing it all, a natural writer with consummate skill at conveying wonder. He belongs to the same select band as Gilbert White and Nan Shepherd, for whom the mere speaking or writing of a place name should invoke their role in chronicling it. Muir wrote much about the High Sierra, the Mountains of California, and Yellowstone and General Grant National Parks, as well as specific observations on their flora and fauna, including essays on the Water Ouzel, the Douglas Squirrel and Wild Sheep. During the spring of 1869, he took the job of a sheepherder on a ranch in the Sierra Nevada as a means to an end to fund his explorations through California's Central Valley. As he explored, his growing familiarity enabled him to become proficient enough to act as a guide, and some of The Yosemite's most famous visitors followed him on his treks, including, among others, the writer Ralph Waldo Emerson. All these experiences were added to Muir's documentation, which culminated in the publication of *The Yosemite* in 1912.

His witness of being present during a night-time earthquake in The Yosemite remains, for me, an encapsulation of the awe the painters in America's 'sublime' movement sought to capture: even a sound recording of the event could not surpass Muir's account. It is pictorial, sonic and poetic, and conveys the complete impression of being at the centre of a giant cataclysmic event. And it begins in stillness:

> It was a calm moonlight night, and no sound was heard for the first minute or so, save low, muffled, underground, bubbling rumblings, and the whispering and rustling of the agitated trees, as if Nature were holding her breath. Then, suddenly, out of the strange silence and strange motion there came a tremendous roar. The Eagle Rock on the south wall, about half a mile up the Valley, gave way, and I saw it falling in thousands of the great boulders I had so long been studying, pouring to the Valley floor in a free curve luminous from friction, making a terrible sublime spectacle – an arc of glowing, passionate fire, fifteen hundred feet in span, as true in form and as serene in beauty as a rainbow in the midst of the stupendous, roaring rock-storm. The sound was so tremendously deep and broad and earnest, the whole earth like a living creature seemed to have at last found a voice and be calling to her sister planets. In trying to tell something of the size of this awful sound it seems to me that if all the thunder of all the storms ever heard were condensed into one roar it would not equal this rock-roar at the birth of a mountain talus. Think, then, of the roar that arose to heaven at the simultaneous birth of all the thousands of ancient canyon-taluses throughout the length and breadth of the Range! (Muir, 462)

I quote this passage at length because it seems to me to contain the essence of a cataclysmic event emerging from stillness, contained within the compass of one person's sense, its connection between what he saw and what he heard, recorded faithfully and exactly. Here is the juxtaposition of a sight 'serene as a rainbow', accompanied by the 'stupendous, roaring rock-storm'. Few can have observed such a spectacle and survived to write such an account. Muir creates an astounding visual image in his words, but it is the sound that seems to reverberate on the very page on which the words are printed. Not only does this continue to exist in the memory, but Muir takes us on his own imaginative journey at the end, suggesting that if this was caused by just *one* mountain reshaping itself, can we conceive what the sound of the creation might have been? In that sense, it is truly Biblical.

Yet equally powerful is a tiny sound placed into a great silence; a single drop of water falling onto a hard surface in a huge resonate space carries all the qualities of a fading bell. Its initial impact shocks the air, and then, the mind follows the duration of its echo into space. Silence in an empty house, shattered by one creaking board, lingers on in the consciousness, haunting us. Hildegard Westerkamp, recording on a side channel of the Bow River in Banff, where the water was frozen in horizontal sheets, layered on top of one another, noted that 'by rubbing small chips of ice on the contours of the large sheets of ice, a very glassy resonance was recorded, ever-changing in its timbre as it encountered different hollow spaces beneath. The sounds had an almost unbearable crispness to them if amplified too much. Because of its incredible clarity the surrounding silence seemed amplified' (Westerkamp in Augaitis and Lande, 91–2). Our responses form partnerships between the material and the sensory; the ear hears, but the mind makes implicative connections. In this sense, a sound to the attentive brain can exceed its own containment.

Enemies of the Event

Jacques Attali wrote: 'For twenty-five centuries, Western knowledge has tried to look upon the world. It has failed to understand that the world is not for the beholding. It is for the hearing. It is not legible, but audible' (Attali, 3). The problem would seem to be that while we have the ability to close our eyes, we cannot do the same to the ears: we have no 'ear-lids'. Thus, the brain is required to make choices of listening, and it does so by focusing itself on attention to certain sounds and disregarding others, while recognising their presence in the soundscape. Within certain environments, this of course becomes more difficult. Robinson's Connemara storms, for instance, or in a noisy club, a factory, a babble of excited voices in a restaurant or simply the hubbub of a loud city street—the thought process that seeks out sounds that may be interesting, unusual, frightening or inspiring becomes weary of the task and can in some cases abandon it. There is even a saying for it: 'I can't hear myself think'. It is rather like trying to have a conversation with an overbearing companion who talks too much and too loudly. In the end, we give up. We may still be in Berger's field, our field of sonic consciousness, but the field is now full of fog. Some of this fog surrounds us through situation and circumstance;

on the other hand, some of it is imposed on us by ourselves, or rather by those who would manipulate us, particularly in the world of retail. Background music in public places has the same effect as background noise; at its most familiar, it is about consumer integration and a levelling of classes and cultures. Beyond that, as Attali has said, 'it is a means of silencing, a concrete example of commodities speaking in place of people' (Attali, 111).

The term 'Muzak' has become a generic one for background music of any kind, but in fact the word has its origins in a commercial company of that name, formed in 1922 to provide telephone music, moving later (in 1940) into the field of the marketing of atmosphere music. Attali, referring to it as 'the music of silencing', points to the practical philosophy behind the construction of Muzak:

> The pieces of music used…are the object of a treatment akin to castration, called "range of intensity limitation," which consists of dulling the tones and volume. They are then…classed by length and type of ensemble, and programmed by a computer into sequences of 13 ½ minutes, which are in turn integrated into completed series of eight hours, before being put on the market. (ibid.)

The confessed aim of such organisations is not to sell music but to sell programming, and we have no choice as we travel in our trains, buses, elevators and taxis to our offices, shops and restaurants, but absorb it. Such is its ubiquity that we do not even notice it much of the time, becoming immune to its cause, but not its effect. Only when, occasionally, the volume in a restaurant grows a little too loud, do we perhaps complain, and once in a while, there may come a tune we actually recognise. To imagine that it is there for our pleasure, however, is mostly a fallacy, although it is certainly true to suggest that through decades of becoming almost submerged in its homogenised 'soup' it has become a kind of non-event in our sonic world which, even if we did not miss it, we would certainly notice its absence were it to be removed. Latterly, copyright agreements have enabled some establishments to play recordings of familiar artists and renditions of songs as background music, thus creating background music that at the same time imposes itself as an event, pushing its way into our consciousness as an unwanted distraction.

It can, of course, work the other way; listening to a 1961 classic recording of the great jazz pianist, Bill Evans, playing at his favourite venue, the Village Vanguard club in New York, I am faintly aware of a hum of conversation with which the music is forced to compete. While the recording is a good one, and my attention is not distracted, I am conscious of its presence. For the most part though, I stay entranced by the subtle music of Evans, his drummer, Paul Motian and his bassist, Scott LeFaro, who, the books tell me, would be killed in a car accident within weeks of this gig. This then is not simply music; it is an event, an account of a temporal happening, existing on several levels at once. It is sound, history, human existence, creativity, art, the capacity to improvise within a situation, the sympathy between minds and a document of something that can never happen again between these three people; had we been in the room that night, we would have been part of a small elite group of witnesses who could say, 'I was there'. Hindsight is one thing; nonetheless, for some of the residents, sipping their cocktails, it would seem that what was going on there on the small stage was truly incidental to their evening. For some perhaps, the Bill Evans Trio was just the background ambience, the mood music, part of consumer integration and not an event in itself at all. This might be a first impression, and it would of course be grossly unfair for those who came for what the musicians gave them. There exists a series of recordings collectively known as *Bill Evans: The Secret Sessions*, which are in effect semi-bootlegs of Evans at the Vanguard, recorded by an obsessive enthusiast for the music called Mike Harris, who taped it between 1966 and 1975. Without the finesse of a professional commercial recording, these tapes convey a sense of presence, of musicians at work within their familiar environment, in short, a feeling of music as event in real time. It is true that many jazz musicians—Dave Brubeck among them—found the recording studio sterile and inhibiting, a place where 'the event' was more difficult to extricate from the imagination. Similarly, Frank Sinatra believed he worked best in the recording studio when he placed himself within the body of the accompanying orchestra, rather than in a sound booth, or worse, as a voice track added at a later point. For them, and clearly for Evans, the intimate and immediate performance setting released something. It was Louis Armstrong who once said words to the effect that jazz is music that never sounds the same twice, and the twice-nightly routine of club performance, or the simulation of the live set, even in a recording studio, makes every rendition of a song a new event which may or not this time touch something sublime. Yet I still cannot help noticing that murmur of voices at the Village Vanguard; do they realise that at this moment, their conversation is taking their attention away from a moment of genius? Or do they not care? Has background noise and background music become so much of a way of life, that when the real things happen, we simply do not notice them? Under the layers, there are other layers. We should be ecologists of sound; as Hildegard Westerkamp says: 'The small, quiet sounds in the natural environment are symbolic of nature's fragility, of those parts that are easily overlooked and trampled' (Westerkamp in Augaitis and Lander, 91). The first thing we hear may not be the most important, and the loudest or most enveloping sound may not be the most significant event after all. We feel our way through sound by way of exact attentions.

Note

1. Street, Seán. *Sound Poetics: Interaction and Personal Identity*. Cham: Palgrave Macmillan, 2017.

3

The Sounds the Ghosts Knew: Imaginative Remembering of the Past

In a Green Shade

On 2 January 1769, the naturalist and priest, Gilbert White, wrote a letter from his home in Selborne, Hampshire, to his friend, Thomas Pennant on the subject of the unique sound of the nightjar, a nocturnal bird that was surrounded by considerable superstition. The bird's 'music' was—and is—unique, a strange churring sound which White describes in his letter as being produced through the windpipe, 'just as cats purr'. He also noted the punctuality of the start of its 'song', 'just at the report of the Portsmouth evening gun, which we can hear when the weather is still'. What is of particular interest to us here is that the distance between Selborne and Portsmouth on the south coast of England is some 29 miles, with several towns and a range of hills known as the South Downs lying between. We have heard of guns from the Western Front being heard in the south-east of England during the First World War, but White's assertion is nevertheless a dramatic example of the capacity for sound to carry across a landscape that today hums and buzzes with noise of all sorts. It is an authentic witness that demonstrates powerfully how sound transmitted across the land in his day without the impediments it now encounters.

© The Author(s) 2019
S. Street, *The Sound inside the Silence*, Palgrave Studies in Sound,
https://doi.org/10.1007/978-981-13-8449-3_3

In a later chapter, we shall examine the nature of imagined sound as contained within works of art, ideas conveyed to the viewer through observation and personal interpretation. While a painting or a sculpture may continue to communicate its message and emotion over centuries, even millennia, sound is, by its very nature, ephemeral, while being at the same time an immediate and vital force; it is at one and the same time, dynamic and evanescent. Because its waves dissipate into a void, we are left with sound that has faded beyond us whenever we seek to personalise the past. The dead leave us their art, their words and their ideas, but until recent history, they have not left us their physical voices, and the places and styles they occupied and adopted have in many cases been so changed, and in some cases, destroyed, that the people who lived before electromagnetic recording seem to us to be dumb, as if we look at one another through a plate glass window through which their world and ours must resort to sign language. In order to 'hear' them and their world, we must become imaginative receivers, as Bruce R. Smith has so well put it, to '"un-air" sounds that have faded into the air's atmosphere' (Smith in Bull and Black, 129). What started Smith on the path to what ultimately became his book, *The Acoustic World of Early Modern England* was 'the thought…that all the sounds that ever occurred still vibrate, however faintly, somewhere in the wild blue yonder…' (ibid., 128).

The products of the microphone and recording technology can take us a limited way back into the tunnel of the past; there then comes a point where we can only imagine how the world sounded before our time in it. That imagination might carry with it the idea that the world was much quieter in previous times, and indeed, in some ways, and in some places, it clearly was. There was no roar of aircraft, no persistent hum of traffic on distant or close motorways, no electronic media thumping or chattering from house or from passing vehicle. On the other hand, localised noise in city environments would have been cacophonous. It is here that we must turn to literature as an aid to the triggering of recognition. Writers of the past like White and Charles Dickens, and poets such as Alfred, Lord Tennyson, Edward Thomas and George Crabbe, help to push open a door into a room that time has silenced. Before we could record with technology, we could evoke through the power of imagination, expression and description. It is this capacity that still speaks, mind to mind, and connects us to how life used to sound. Crabbe in particular, through his long narrative poems, *The Borough*, *The Parish Register* and *The Village* written during the late eighteenth and early nineteenth centuries, provides a voice that we may recognise, while noting how passing time spreads a gauze of strangeness brought about by the changes in the intervening years over the pictures they provide. In a world where darkness and shadow held much greater sway, where every sound that broke in on silence was noteworthy, if not sometimes alarming, the human capacity for listening was tuned in a more subtle and animalistic way than we with our desensitised sonic abilities can understand. Today, we must make conscious efforts to develop 'active listening' that to our ancestors was a key and instinctive part of understanding the world, and indeed, surviving within it. We shall use the written text here as a recorder, and the reading eye as a listening device. Above all, we shall consider how yesterday became today through clues left in the audio vestiges on the walls of time. Amateur poets are often the most eloquent in this respect; John Warren, 3rd Baron de Tabley, for example, gives us sound pictures of farm life before the coming of machines, as in the start of this description of a rural evening. He was writing in the second half of the nineteenth century, but the images he evokes are redolent of a bucolic scene that seems almost timeless.

> The whip cracks on the plough-team's flank,
> The thresher's flail beats duller.
> The round of day has warmed a bank
> Of cloud to primrose colour.
>
> The dairy girls cry home the kine,
> The kine in answer lowing;
> And rough-haired louts with sleepy shouts,
> Keep crows whence seed is growing.
>
> The creaking wain, brushed through the lane
> Hangs straw on hedges narrow;
> And smoothly cleaves the soughing plough,
> And harsher grinds the harrow.

> Comes, from the road-side inn caught up,
> A brawl of crowded laughter,
> Thro' falling brooks and cawing rooks
> And a fiddle scrambling after. (De Tabley in Carr, 60)

Simple as the writing is—almost doggerel—nevertheless, the picture it offers is a vivid depiction a warm evening in early summer sometime in the nineteenth century. It is all slow, drowsy sound, thrown into relief by the sense of rural stillness that surrounds it. This was the kind of scene, unchanged for virtually centuries, that was, even as de Tabley was recording it in verse (he was born in 1835, and died in 1895), in the process of being overtaken as industry moved its machinery into the landscape.

That, at least, is the impression we may have of the effects of the industrial revolution. The years before the First World War have frequently been seen through the wistful rose-tinted glasses of nostalgia—a golden gentle age of quiet, arcadian innocence that was destroyed by the subsequent industrial cacophony of conflict, urban noise and squalor. Yet we should also remember that even as de Tabley was penning his verses about the idyll of country life, perhaps over the next hill, steam hammers and furnaces were roaring. Imagined and sometimes idealised sound has always played a major part in the construct of past sonic worlds, because we have no direct witness other than the written word, and this chapter seeks to explore in particular some of the conundrums associated with the period 1900–1914, as well as its incipient nationalism and idealism, through the memory of sound, music and literature. We shall also seek to delve further back and listen.

Fire and Ice

We hear them first in accounts of major events, the sounds of war, national celebration and disaster, as in the staple witness of writers such as John Evelyn and Samuel Pepys. Often quoted as it is, Pepys's account of the Great Fire of London across five terrible days between the second and the sixth of September 1666 never loses its power to shock and sadden us, and such is his descriptive powers that our own sensory imagination time-travels to meet him in a city which was, even as he watched it, vanishing into flame. The significance of the moment is that it is the violent signal and metaphor for the change which, although usually it occurs more slowly, nevertheless is constantly either destroying or burying the past beneath layers of time. Reading Pepys, we become aware of another element in how our senses intercommunicate. One of the strengths of radio—which we shall explore further in a subsequent chapter—is that because we hear, we mentally see. At the same time, vivid written passages communicate through the eye to the brain, creating both imaginative images and 'silent' sound. Take these extracts from Pepys's *Diary*, drawn from across several days:

> So I made myself ready presently, and walked to the Tower…and there did I see the houses at that end of the bridge all on fire, and an infinite great fire on this and the other side the end of the bridge….Everybody endeavouring to remove their goods, and flinging into the river…and the wind mighty high and driving it into the City; and everything, after so long a drought, proving combustible, even the very stones of churches…And between churches and houses, as far as we could see up the hill to the City, in a most horrid, malicious, bloody flame, not like the flame of an ordinary fire…We stayed till, it being darkish, we saw the fire as only one entire arch of fire from this to the other side of the bridge, and in a bow up the hill for an arch of above a mile long: it made me weep to see it. The churches, houses, and all on fire, and flaming all at once; and a horrid noise the flames made, and the cracking of houses at their ruin… and Lord! To see how the streets and the highways are crowded with people running and riding, and getting of carts at any rate to fetch away things… (Pepys, 178–83)

Even where Pepys does not name sound, we hear it as we see the pictures he paints in words: the rushing of the blaze, the cries of people and animals, the splitting of wood and the falling of stone. Likewise, John Evelyn's account of the same event complements Pepys in its graphic description of the destruction of the city and the mind-numbing cacophony as walls fell around him: 'the stones of St. Paul's flew like grenades, and the lead melted down the streets in a stream'. It seemed to Evelyn to be a glimpse of Armageddon:

The noise, crackling and thunder of the impetuous flames, the shrieking of women and children, the hurry of people, and the fall of towers, houses and churches, was like a hideous storm…Thus I left it this afternoon burning, a resemblance of Sodom, or the Last Day. London was, but is no more. (Evelyn, 154)

Sight and sound intertwine, ebb and flow, like tides; an image flashes before us, deposits a sound in our mind and then is overtaken by another wave, which does the same; such is the power of this interaction that as the visual imagination is triggered, so is the auditory, and a soundscape, full of implied perspective, plays in a kind of mental 'surround sound' inside us. In terms of a commentator's engagement with what Pepys and Evelyn are witnessing, such descriptions are reminiscent of the famous 'live' radio account by Herb Morrison of the destruction of the airship, 'Hindenburg' at Lakehurst, New Jersey on 6 May, 1937, during which at one point, overcome with emotion, Morrison cries out, 'Oh, the humanity!'[1] We shall encounter Morrison's New Jersey broadcast again later in this book.

While the great cataclysmic events of history open doors onto moments in past times, it is in the every day, in the world of routine and utility as it happened around people, that we most often find the eloquence and poignancy of lost lives and times, much as looking at a street scene in a photograph, on a day when nothing extraordinary was going on, reveals more to us about what is lost and what remains. It is in such images and accounts that our kinship and empathy with our ancestors may be sensed most profoundly. John Evelyn wrote of the severe winter of 1683/1684, when famously the Thames froze over sufficiently for commerce and entertainment to take place on its ice. History books may tell us of the circumstances, but accounts such as Evelyn's convey to us almost Brueghel-like pictures of the event in human terms, which in turn play a soundtrack of what it meant to be in London at the time:

January 9th, 1684. I went across the Thames upon the ice, which was now become so incredibly thick as to bear not only whole streets of booths, in which they roasted meat and had divers shops of wares as in a town, but also coaches and carts and horses which passed over… (Evelyn, 195)

By the 24th of the month, he noted that the frost was still continuing, and was, if anything, more severe, and the novelty of the Thames 'planted with booths in formal streets, as in a city or continual fair', lured him back:

> All sorts of trades and shops were furnished and full of commodities…-
> Coaches now plied from Westminster to the Temple, and from several other stairs to and fro as in the streets. There was likewise bull-baiting, horse and coach races, puppet plays and interludes, cooks and tippling – and lewder places – so that it seemed to be a bacchanalia, triumph, or carnival, on the water. (ibid.)

Eighteen years after witnessing the destruction of London by fire, and the decimation of its populace, Evelyn gives us his imaginative 'recording' of the sound of some of its surviving citizens at play.

Figures in a Landscape

We have met the poet Edward Thomas already on this journey. Now, we come across him again, walking through a village near Trowbridge in Wiltshire. It is an evening in the spring of 1913, about 9.30pm:

> A few stars penetrated the soft sky; a few lights shone on earth, from a distant farm seen through a gap in the cottages. Single and in groups, separated by gardens or bits of orchard, the cottages were vaguely discernible: here and there a yellow window square gave out a feeling of home, tranquillity, security. Nearly all were silent. Ordinary speech was not to be heard, but from one house came the sounds of a harmonium being played and a voice singing a hymn, both faintly. A dog barked far off. After an interval a gate fell-to lightly. Nobody was on the road. (Thomas 2016, 164–5)

Walking on, he hears a dog growl at his footsteps. At a gate he stops, feeling the silence. 'A train whistling two miles away seemed as remote as the stars. The noise could not overleap the boundaries of that silence'. The whole experience seems to have had an almost trance-like effect on him, darkness punctuated by beads of light and sound: 'I felt that I could walk on thus, sipping the evening silence and solitude, endlessly'. Arriving at his lodgings for the night, he finds himself returned to the world: 'I entered, blinked at the light, and by laughing at something, said with the intention of being laughed at, I swiftly again naturalised myself' (ibid.).

Edward Thomas's book, *In Pursuit of Spring*, from which this passage comes, was written in 1913, when he was on the cusp of discovering himself as a poet, and is a sourcebook of the sounds of its time, drawn from particular environments, just before the First World War. Here, he gives us a sonic archetype that fits our imagined perception of what the (for the most part rural) world must have sounded like at the time, heightened by our hindsight, a stillness about to be shattered. In 1914, the same year that saw the publication of his book, James Joyce published *Dubliners*, a book of stories written, largely at a distance from his native city; thus reading them today, we have both a detachment of time and through the author's eyes and ears, a self-imposed but intense 'over-the-shoulder' observation of place, at a key moment in its history. In his story, 'Araby', we have a boy walking through an urban street, full of sound and sensations:

> On Saturday evenings when my aunt went marketing I had to go to carry some of the parcels. We walked through the flaring streets, jostled by drunken men and bargaining women, amid the curses of labourers, the shrill litanies of shop-boys who stood on guard by the barrels of pigs' cheeks, the nasal chanting of street-singers, who sang a *come-all-you* about O'Donovan Rossa[2], or a ballad about the troubles in our native land. These noises converged as a single sensation for me. (Joyce, 18)

Joyce and Thomas were writing at the same time, and for the fancy, we may imagine the latter's street scene happening at the same moment as the village soundscape of the former. When considering the sounds of the past, it is vital to keep in mind how various they were. There have always been crowds and stillness, large and small communities, noisy and quiet environments, happening in their designated places simultaneously. No one sound can therefore definitively depict or represent a time or period in history. Equally, we need to remember that Thomas's village account was written by him as a stranger, a visitor—his senses sharpened by lack of familiarity; Joyce, on the other hand, was a part of the world he described, yet he too 'heard' the scene he described through the heightened filter of memory. All that apart, both men wrote what they knew and what they heard at the time, and it is a temporal place to which we have no way of returning, save through what their texts have given us; we must trust their witness.

Edward Thomas's most famous poem 'Adlestrop' is a perfect expression of the growing awareness of sounds as all else in the world steps back and our emotional and psychic antennae tune in. It prints on the memory of the reader just as the sounds of the moment printed on his own: a hiatus of rural stillness in the summer of 1913, when his train stopped unexpectedly at a village station, somewhere in Gloucestershire. I have discussed Thomas's poem and quoted it in a previous book (*Sound at the Edge of Perception*). Suffice to say here that, at the time of writing the poem, he was travelling to spend time with a group of fellow poets who had established what they hoped would be a fraternity of like minds in Gloucestershire. This fellowship has subsequently become known as 'The Dymock Poets', centring as their activities did, around the small village of that name south of Ledbury.

The presence of sound events in the Dymock countryside transmits through much of the writing of the poets who came there. Some of them were, after all, what we might term, 'townies' for whom the stillness and relative stability of the landscape by its very comparison with their more familiar urban terrain accentuated to their ears its own special qualities, in which every sound became significant. The presence of the American Robert Frost and his 'Sound of Sense' theories would also have been a great encouragement to awareness and discussion, but there is a feeling in the writing of most of the poets that a new and poignant consciousness of the subtleties of the sonic world gained a response in their work, from the ambience that met them as they opened their cottage doors to the transient presence of voices. We may make much of the fact that their experiences came just before the brutalising noise of a new war that to some extent erased the fragile music of social continuity and desensitised the capacity for active listening in human beings. Yet as we have seen, mechanisation, the internal combustion engine, aircraft and radio, would soon impose themselves in unimagined ways; already the dull roaring of towns and cities was bleeding into field and hedgerow, beginning the change not only of the sound of the world, but in our ability to hear the detail of the small sounds of nature and the events that had punctuated stillness since Gilbert White's day and beyond. It was to bring a blunting of sensitivity that diminished centuries of awareness. In his introduction to an edition of White's *Selborne*, Richard Jefferies, a hero to Edward Thomas, wrote:

> Anyone who desires to see some of the things that this man saw…cannot do better than fix himself in some pleasant spot, and work there in absolute quietness for as many days as possible. For it is in this quietness that the invisible becomes visible. The vacant field gradually grows full of living things. In the hedges unsuspected birds come to the surface of the green leaf to take breath. Over the pond brilliantly coloured insects float to and fro, and the fish that never seem to move from the dark depths do move and do come up in sight. (Jefferies in White, x–xi)

Sound as we know can be fragile, or it can be ferocious; in the years up to the First World War, cities could be, then as now, aggressive places. While Richard Jefferies was listening to stillness on the Wiltshire Downs, Charles Dickens in *David Copperfield* could write of a London street as containing the 'bawling, splashing, link-lighted, umbrella struggling, hackney-coach-jostling, pattern-linking, muddy, miserable world' (Dickens 1985, 344). Two centuries earlier, city sounds would have been different, but perhaps still recognisable to the characters in Dickens's novels. Take this scene, imagined by the American sound designer David Sonnenschein for example: 'In the Paris of the 17th century, reports of noise included shouting, carts, horses, bells, artisans at work, etc. All these sounds came from specific directions and were very impulsive, rather than long constant tones, with few low frequencies' (Sonnenschein, 96). Noisy it may have been, but there is a sense here of the human, a feeling of heard events happening rather than being consumed in an amorphous din. Proust wrote of vendors' street cries filtering through bedroom windows in the morning; there was an identity to urban activity. It was also utilitarian; the sounds of the city were not ornamental but functional. The noise, in particular, of late Victorian and Edwardian cities would have been made up of metal-wheeled carts and carriages, pulled by horses over cobbled streets, vying with the new technology of trams, motor-buses and cars, all raising the prevailing sound to mind-numbing levels, added to which would have been the shouting of voices, made louder by the effort of being heard over everything else, and so contributing to the overall sound pollution of the early twentieth century.

Today, we add to the sounds around us by personalising our own in order to either draw attention to ourselves, or to seek confirmation that we do still retain an identity as an individual in the crowd. The sound itself too, as Sonnenschein reminds us, has changed:

> In a postmodern environment, the sound may also be omnidirectional, but higher pitched…electronic rather than mechanical. There is a certain sterility and predictability to [our situation], but humans have given character to their ever-shrinking, attention demanding machines – like musically ringing cell phones, for example. (ibid., 97)

In coming years, city streets around the world will undergo another radical change with the introduction and ultimate proliferation of electric vehicles; this in turn raises an interesting question as to the safety factors that may be required, particularly in the early years, when a public, used to the warning sound of an approaching bus or car, may need a device or sound on the vehicle to draw their attention to its presence. So used to noise have we become, that we may feel displaced and threatened without it within certain environments.

Stepping out of a front door in the city in past times meant encountering the noisome noise of horse-drawn traffic, and as the sound evolved, it became part of a metaphor. Writing of Edwardian Britain, Jose Harris has pointed out that 'many features of pre-war society – changes in the structure of family life, the emergence of the labour movement, the challenge of feminism, the investigation of poverty, the rise of aesthetic modernism, and the growth of moral and religious uncertainty – seem to anticipate concerns of the later twentieth century' (Harris, 2). This was built on the growing mechanised hubbub of the industrial revolution, so that, while the violence that generated the cacophony of war in 1914 shocked the twentieth century into a new start, it was also the cultural crescendo of sonic assault that had been growing for decades. In some ways, we may consider the years 1900–1914 to be a continuation of the nineteenth century, but in another sense, the world we have come to call 'modern' was already forming itself. By the mid-nineteenth century, the economic climate was changing at a bewildering rate, and the shift in the structure of work may be seen by considering that, while in 1811, agriculture was the most important industry in Britain, with over a third share of the national income; by 1861, that had reduced by almost 50%, and by 1901, it was down to 6%. As with the economy, so with political power; by the outbreak of the First World War, this had passed irrevocably to the industrial and commercial middle classes. In the process, the subtle sounds of a rural life began to sink under new noise as the countryside became less a place of industry, and (more in terms of the economy at least) a kind of 'park', seen as a place of refuge and leisure from the increasingly pervasive noises of modern civilisation. James Mansell has put it well:

> Noise was not just representative of the modern; it *was* modernity manifested in audible form. Discussing noise allowed commentators to express what it *felt* like to be living in modern times. Noise was clamorous, all-enveloping, and unpredictable. Discussions of the age of noise constituted a conscious engagement with the politics of modernity in which the modern was not primarily a set of ideas, institutions, or practices, but rather the sensed experience of an unnerving atmosphere. (Mansell, 1)

Within the ubiquitous sound, there remained an ability inherited from the previous generation for attentiveness. Some could utilise it as a tool of the trade, as with the fictional stonemason, 'Stony' Durdles, in Dickens' *The Mystery of Edwin Drood*:

> "Now, look'ee here. You pitch your note, don't you, Mr Jasper?'
> "Yes."
> "So I sound for mine. I take my hammer, and I tap…tap, tap, tap. Solid! I go on tapping. Solid still! Tap again. Hallao! Hollow! Tap again, persevering. Solid in hollow! Tap, tap, tap, to try it better. Solid in hollow; and inside solid, hollow again! There you are! Old 'un crumbled away in stone coffin, in vault!" (Dickens 1972, 35–6)

As John M. Picker reminds us in his study of nineteenth-century social sound, 'the Victorian soundscape was so varied and vast as to be too much for one pair of ears to apprehend…this was a period of unprecedented amplification, unheard-of loudness…alive with the screech and roar of the railway, and the clang of industry, and with the crackle and squawk of acoustic vibrations on wax and wire – yet alive as well with the performances of the literary figures who struggled to hear and be heard above or through all of this' (Picker, 4).

Blending with attempts to preserve sound through words was the emergent technology that would assist and later diminish it, such as the microphone, loudspeaker and the phonograph. David Hughes created his prototype microphone, a forerunner of numerous subsequent carbon versions, as early as 1878, and from 1879 to 1886, experimented with wireless signals up to 500 yards. A year earlier, Edison had invented the phonograph, and a year prior to that Alexander Graham Bell patented his first electric loudspeaker, as part of his 1876 telephone. In 1877, Ernst Siemens improved the technology further; thus, within a decade, the noise of the world became transmittable, recordable and ubiquitous. It is perhaps no surprise that one of the most sound-conscious books of the era, *Daniel Deronda* by George Eliot, was published in 1876. As Picker has eloquently expressed it, 'The anxiety in Eliot's final novel…is the burden of the stethoscopic age:[3] that of hearing too much and too well. *Daniel Deronda* is, among other things, about choice: in this era of close listening, when there is so much new to hear, to what and to whom should one listen?' (Picker in Morat, 29).

Eliot was familiar with the work of the inventor of the stethoscope, Rene Laennec, and he is actually mentioned in her previous novel, *Middlemarch*. She also knew of the work of the German physician and physicist, Hermann von Helmholtz, who had been exploring sound since 1856, and in 1857, he delivered a key lecture, 'The Physiological Causes of Harmony in Music'. There is no doubt that Eliot and her common-law husband, George Henry Lewes, knew this; indeed, they owned copies of Helmholtz's writing in German and French, and Eliot herself noted that she had been reading his theories of music. In the introduction to her 1866 novel, *Felix Holt*, she had written:

…there is much pain that is noiseless; and the vibrations that make human agonies are often a mere whisper in the roar of hurrying existence. There are glances of hatred that stab and raise no cry of murder; robberies that leave man or woman for ever beggared of peace and joy, yet kept secret by the sufferer – committed to no sound except that of low moans in the night…Many an inherited sorrow that has marred a life has been breathed into no human ear. (Eliot, 11)

This passage turns a key in the door of time; most of all, we hear the suppressed sounds of Victorian emotions, the subjugation of women, suffered and resented in silence against the growing cacophony of an increasing urban and industrialised world, in which even the sound of thought itself seems amplified, and new technologies probe and interrupt us and trouble our quietness. Eliot is aware that sound is vibration:

Helmholtz's theory of sympathetic resonance, which explained how the ear as a kind of "nervous piano" was able to perceive musical notes, itself sympathetically resonated with Eliot's aesthetic project to dramatise the varieties of close listening through which her characters, and by extension her readers, develop compassion and affinity. (Picker in Morat, 29)

In the meantime, many city dwellers had sought to combat, unsuccessfully, the sounds around them: '…Besieged by street noises that interrupted his writing, Thomas Carlyle invested in a plan to construct a sound-proof study at the top of his house. But once it was finished, he found it difficult to work there, claiming the shock of stray sounds had become even worse than before' (ibid., 6). All this in the context of industrial noise, which infiltrated the air with the pollution from factory chimneys, just as the physical structures spread across landscapes. Yet although ideas of the industrial revolution may conjure Blake's 'satanic mills', or Miltonic images akin to the apocalyptic art of John Martin, for many there was a heroism, a kind of sublime power in the furnace and the hammer. In 'Steam at Sheffield', the Yorkshire poet Ebenezer Elliott, born in 1781, wrote of the sound of the factories as 'beneficent thunder':

Oh, there is glorious harmony in this
Tempestuous music of the giant, Steam,
Commingling growth, and road and stamp and hiss,
With flame and darkness! Like a Cyclops' dream… (Elliott, quoted in Drabble, 201)

The blend of industry and nature as one impinged on the other provided visual and auditory spectaculars for travellers, particularly when seen from above. Arthur Young, writing in 1776 in *Annals of Agriculture, and Oher Useful Arts*, found it all 'horribly sublime', writing that:

> Coalbrook Dale itself is a very romantic spot, it is a winding glen between two immense hills which break into various forms, and all thickly covered with wood, forming the most beautiful sheets of hanging wood. Indeed too beautiful to be much in unison with that variety of horrors at the bottom: the noise of the forges, mills and all their vast machinery, the flames bursting from the furnaces with the burning of the coal and the smoak [sic] of the lime kilns, are altogether sublime... (Young, quoted ibid., 194)

The growth of industry and its impact on landscape through the seventeenth and eighteenth centuries was a counterpoint to the thinking of a number of philosophers and poets. It was in 1757 that Edmund Burke discussed the idea that our aesthetic responses to the world are experienced as pure emotional arousal, unencumbered by intellectual considerations. Certainly for many, at a time when long-term environmental issues tended not to be uppermost, and the excitement of the spectacle of the dark temples to industry possessed a certain awe and intoxication, Burke echoed new and novel visual and auditory images. Indeed, these great flaming factories and mills in their valleys had a Miltonic grandeur about them and appeared to the inspired onlooker as almost dramatised extensions of the natural world itself, and Burke had written, 'the passion caused by the great and sublime in *nature*...is astonishment; and astonishment is that state of the soul, in which all its motions are suspended, with some degree of horror' (Burke, 47). For Burke, horror, terror and astonishment were terms which we might bring together under the banner of 'awe'. At certain moments, and observing certain phenomena, this fills the mind '...so entirely...with its object, that it cannot entertain any other, nor, by consequence reason on that subject which employs it. Hence arises the great power of the sublime, that...anticipates our reasonings, and hurries us on by an irresistible force. Astonishment...is the effect of the sublime in its highest degree' (ibid.). John Muir, witnessing The Yosemite earthquake in the previous chapter, experienced Burke's sublime awe in its purest form, and we see manifestations of these ideas of the sublime as shown through industrial art as the inspiration for many painters of the period and the apocalyptic artwork of the aforementioned John Martin. The imagery of John Milton's epic poem, *Paradise Lost*, was here witnessed in human terms, and for some it was heroic.

Nevertheless, it is clearly little wonder that in contrast to this, even while it was viewed as positive and exciting by some, aspects of the rural life still offered the sense of a more stable and peaceful escape, and likewise, it was hardly surprising that city dwellers—if they had the funds and time to do so—would seek their own 'escape to the country' whenever they could, or at least imagine it through the written word. Consequently, a glut of books such as Eleanor Haydn's *Travels Round Our Village*, published in 1905, fed the dream that created a lucrative literary market:

> It is good in these days of bustle and strife, to drift for a while into some quiet backwater – such as may yet be found in rural England – which the tide of progress stirs but just enough to avert stagnation; where old-world customs and archaic forms of speech still linger and where men go about their daily tasks in a spirit of serene leisureliness, therein copying Nature who never hurries. Of such a sequestered corner, its humours, its homely comedies and simple pathos would I write. (Haydn, 9)

Yet the very railway culture that gave many of Haydn's readers their direct access to this rural idyll obliterated landscapes and scarred the air as it desecrated the landscape. In *Dombey and Son*, Charles Dickens provides us with a graphic sound picture of steam locomotion:

> Night and day the conquering engines rumbled at their distant work, or, advancing smoothly to their journey's end, and gliding like tame dragons into the allocated corners grooved out to the inch for their reception, stood bubbling and trembling there, making the walls quake, as if they were dilating with secret knowledge of great powers yet unsuspected in them, and strong purposes not yet achieved. (Dickens 1974, 218–9)

A further irony, as an aside, could be that these very steaming monsters were responsible for transporting Dickens to the audiences that would clamour to hear him read such passages. Not for the last time would the general public desire a new dream, whilst clinging to an archaic one. As late as 1926—the very year of the General Strike—Stanley Baldwin was to write a kind of manifesto for nostalgic values as a basis for nationalism and patriotism:

> The sounds of England, the tinkle of the hammer on the anvil in the country smithy, the corncrake on a dewy morning, the sound of the scythe against whetstone, and the sight of a plough team coming over the brow of a hill, the sight that has been seen in England since England was a land… (Baldwin, 7)

The political and cultural exploitation of the patriotic in the service of nationalism within the UK has often employed the bucolic and rural rather than the industrial and urban for its effect, just as it has usually turned to the evocation of 'Dream England' rather than 'Britain'. Needless to say, the romanticised sounds and images of rural world upon which Miss Haydn and Mr. Baldwin concentrated would not have been appreciated by the poorest and most disadvantaged members of rural society. Many endured conditions and a level of poverty that necessitated generations to migrate from the west country, for example, to far places such as Newfoundland, in desperate attempts to better a life that in some cases was lived on almost equal terms with cattle. Being poor in the countryside often meant constant labour for adults and children alike, and the demands took a toll on both individuals and their communities. Richard Jefferies chronicled the hardships too, and noted that then, as now, the gap between prosperity and abject penury was extreme, while being at the same time affected by a fine balance that was subject to the vicissitudes of circumstance and societal change:

> When the windmill was new, Peter's forefathers had been, for village people, well off…[Now] to see him crawling along the road by his load of flints, stooping forward, hands in pocket, and then to glance at the distant windmill, likewise broken down, the roof open, and the rain and the winds rushing through it, was a pitiful spectacle. (Jefferies 1987, 48)

Nonetheless, the largely urban-fed dream of arcadian 'England' persists in some forms to this day, through the recording of sounds on the printed page and their transmission as events to the imagination. As Alain Corbin reminds us, 'the resurrection of a lost world is achieved by grasping whatever there is at its heart that today strikes us as most unusual' (Corbin, 308).

Vox Populi

The composer Benjamin Britten, living in Aldeburgh in Suffolk, heard the same town church bells as the poet George Crabbe, who was born there. Crabbe lived from 1754 to 1832, and we hear those bells in one of the 'Sea Interludes' to Britten's opera based on Crabbe's work, *Peter Grimes*. When Crabbe was 26 in 1780, he travelled to London, and it is from that time onwards that we associate the great narrative poems with which he made his name, notably *The Village* (1783) and later *The Borough* (1810). Written largely in heroic couplets, Crabbe's poetry spoke of ordinary people in real-life situations, a radical characteristic in a poetic world that would see the emergence of such writers as Byron, Shelley and Keats. In *The Borough*, he gives us a vivid soundscape of a busy East Anglian port at work; it is almost possible to feel the bite of the chill salty wind off the North Sea:

> Yon is our Quay! those smaller boys from town,
> Its various ware, for country use, bring down;
> Those laden waggons, in return, impart
> The country-produce to the city mart;
> Hark! to the clamour in that miry road,
> Bounded and narrow'd by yon vessel's load;
> The lumbering wealth she empties round the place,
> Package, and parcel, hogshead, chest, and case:
> While the loud seaman and the angry hind,
> Mingling in business, bellow in the wind. (Crabbe, 93)

The scene is full of jostle and the heft of a working quayside, loading, unloading, storing in warehouses and above all, the voices. The most vibrant and telling source of sound in any community must indeed be those of its humanity, the speech patterns, dialects and tropes that identify a place and a time; in terms of the past, we must be grateful to sources that committed to the written word the evidence of such things as vendors' street cries and songs. Ben Jonson gives us pictures of urban life, particularly in its raucous merry-making mode through a number of his works, notably his play, *Bartholomew Fair*. For Jonson, language in all its variety was what formed the sounds of the streets, celebrating 'the sea of voiced city sounds that include the loud chatter of guests, the noises of artisan work and proto-industries, the greater access of women to forms of self-expression, as well as the diverse and often confusing forms of patterned speech, which blend voice, song, ringing, shouting, and a variety of dialects and "professional" accents. Such soundings create a sonic carnival that takes the form of numerous "vaporous" games, pranks, and deliberate investment in the art of noise-making...' (Stanev, 6).

According to Andrew Tuer, in his *Old London Street Cries*, originally published in 1885, the Fair of Saint Bartholomew, depicted by contemporary prints as well as in literature, was, although religious in its origins, a typically raucous affair; by 1721, it was in addition fuelled by alcohol 'in which the lower orders were then accustomed to indulge, unfettered by licence or excise'. He goes on: '...Another woman has charge of a barrow laden with pears as big as pumpkins; and a couple of oyster-women, whose wares are on the same gigantic scale, are evidently engaged in a hot wrangle' (Tuer, 40). Also a regular attraction was 'Fawkes, the famous conjuror', as were ballad singers and the sellers of songs, all of which would have added to the din:

> Let not the ballad singer's shrilling strain
> Amid the swarm they listening ear detain:
> Guard well thy pocket, for these sirens stand
> To aid the labours of the diving hand;
> Confederate in the cheat, they draw the throng,
> And Cambric handkerchiefs reward the song. (ibid., 40–1)

Tuer notes with irony that in a print of Bartholomew Fair made in 1739, 'a ballad singer is roaring out a *caveat against cut purses* whilst a pickpocket is operating on one of his audience' (ibid.). The city streets would have been alive with the voices of salesmen and saleswomen, often attracting custom through rhyme, for example:

> Young gentlemen attend my cry,
> And bring forth all your knives;
> The barbers razors too I grind;
> Bring out your scissors, wives.
> …
> With mutton we nice turnips eat;
> Beef and carrots never cloy;
> Cabbage comes up with summer meat,
> With winter nice savoy.
> …
> Holloway cheese cakes!
> Large silver eels, a groat a pound, live eels!
> Any new river water, water here?
> Buy a rope of onions, oh?

Or:

> Buy a goose?
> Any bellows to mend?
> Who's for a mutton pie, or an eel pie?
> Who buys my roasting jacks?
> Sand, ho! Buy my nice white sand, ho! (ibid., 78–80)

Among the earliest written records of street cries is a verse from the pen of John Lydgate (1370–1451), a prolific Benedictine monk-poet from Bury St. Edmunds. It takes the form of a somewhat rueful account of the financial inability to indulge in retail, under the title, 'London Lyckpenny' or 'Lack Penny':

> Then unto London I dyd me hye,
> Of all the land it beareth the pryse:
> Hot pescodes, one began to crye,
> Strabery rype, and cherryes in the ryes[4];
> One bad me come nere and by some spyce,
> Peper and safforne they gan me bede,

But for lack of money I might not spede.

Then to the Chepe I began me drawne,
Where mutch people I saw for to stande;
One spred me velvet, sylke and lawn,
Another he taketh me by the hande,
"Here is Parys thred, the finest in the land[5];"
I never was used to such things indeed,
And wanting money I might not spede.

Then went I forth by London stone,
Throughout all Canwyke[6] Streete;
Drapers mutch cloth me offred anone,
Then comes me one cryed hot shepes feete;
One cryde makerell, ryster grene,[7] an other gan greete;
One bad me by a hood to cover my head,
But for want of mony I might not be sped.

Then I hyed me into Est-Chepe;
One cryes rybbs of befe, and many a pie
& c. (ibid., 4–5)

We may imagine the noise, the range of voices, dialects, volumes and pitches. Perhaps the nearest we might come today to replicating the sound of such street trade would be in the markets of some Eastern cities and towns. In the early seventeenth century, the English dramatist and pamphleteer Thomas Dekker published a number of non-fiction—or semi-fictionalised—accounts of life in the darker and poorer districts of London. In these, he paints graphic pictures of the social conditions of the time, complete with a strong sense of its sound. Among these, in 1606, Dekker wrote a pamphlet called *The Seven Deadly Sins of London* that gives us an image of daytime London, disabusing anyone who might feel that cities were quieter then: '…Carts and coaches make such a thundering as if the world ran upon wheels…besides, hammers are beating in one place, tube hooping in another, pots clinking in a third' (Dekker, 26).

It is clear that many of the sounds the ghosts of our cities heard when they walked their streets were the sounds of themselves: street activity, entertainment, the cries of vendors and the shouts of passing travellers. In Paris, between the 1820s and after the 1848 revolution, an annual event occurred below Belleville, where the Rue du Faubourg-du-Temple crosses the Canal Saint-Martin, in the form of a parade to mark the end of Carnival in the early morning of Ash Wednesday. It was known as the descente de la Courtille, and a contemporary account by one Privat d'Anglemont conveys the atmosphere, the danger, latent violence, excitement and above all, the sound of this happening:

> Ah, the descent from the Courtille, that was a real bacchanalia of the French people! What a crowd, what confusion! What cries, what noise! Pyramids of men and women clinging to carriages, hurling abuse at each other across the street, a whole city in the street…we might say, without exaggeration, that *tout Paris* was there. Everyone said: "It's monstrous, depraved." But the most refined society, duchesses in domino masks and short-skirted women of easy virtue in their dishevelled finery, courtesans dressed up as bold fishwives, bourgeois as peasants or Swiss milkmaids, hastened at four in the morning to leave the salons of the Opera, the subscription balls, the theatres…to make their way there. There was no good Carnival without a noisy descent from the Courtille…People spilled out of the cheap dance halls, and were everywhere, even on the rooftops; all you could see were heads, all shouting, crying out, splashing each other with wine. Carriages arrived filled with masked figures, and took three hours to get from the boulevard to the *barrière*…People bawled at each other from carriage to carriage, from house windows to carriages, from the street to the windows, each group had its especially loudmouthed character, a kind of rasping corncrake with lungs of steel, whose job was to respond to everyone else. (Quoted by Hazan, 126–7)

Such intensity might have felt to such an onlooker to have the capacity to print itself onto the very fabric of the place, yet today, the sounds exist only in accounts like this, and a few graphic illustrations. The city continues, but the ghosts are gone, and it becomes the role of the imagination to hear their echoes. We return to the sound of bells, the most frequent and lasting connections that persist across time and space, both the sound of a place itself and redolent as they are of the presence of previous generations that have witnessed their tolling across time. In Venice or Florence, the visiting tourist of today will hear their sound as they move around the city as almost a kind of audio décor, a sonic wallpaper that colours the impression with romance. Yet just as with the stations of the day tolled by village bells, and the messages contained in their various peels, we have lost much of the knowledge that made them significant in the past. The importance of bells in the Florence of the Renaissance was not that of abstract sounds spread as points in time across the city; they constituted an interlocking and carefully choreographed dialogue between legislative, judicial and spiritual/sacred institutions. In the preindustrial city, these would have been the loudest sounds its citizens heard, and they were a punctuation to the day that signified a time pattern containing calls to eat, sleep, pray, attend meetings, celebrations and even executions, thus forming a unifying signalling system and as it were encircling the urban community with sound. Were the bells to cease the regular pattern of their chiming, the populace would have been instantly aware and disorientated. As Niall Atkinson wrote, 'pre-modern European communities were, almost without exception, intimately tied to their bells, which in turn bound them to their urban environment. Bells lay at the intersection of popular expressions of communal identities and the disciplinary control of space by the state. As a result, tracing their echoes within urban history is critical to understanding the spatial construction of social relations in urban communities' (Atkinson, 70). Much has altered in the sonic world over the centuries, but in many cases, the sounds the ghosts heard are actually still there: it is we who have changed in our capacity to hear them and to 'read' them. There are still places where the sound of past times continues to surround us, if we but pause and listen in the right way.

Notes

1. Morrison, Herbert, Commentary on the Hindenburg crash, Lakehurst, New Jersey, 6 May 1937. Morrison was making a routine broadcast to experiment with new portable recording equipment. The full commentary may be heard on the CD, *The Aviators*. Pearl/Pavilion Records, index: PAST CD 9760.
2. Irish street songs often began with the summons 'Come all ye…' O'Donovan Rossa refers to Jeremiah O'Donovan (1831–1915), a militant

nationalist and folk hero, known as 'Dynamite Rossa', who underwent imprisonment and exile.

3. The stethoscope was invented in 1816 by René Laennec at the Necker-Enfants Malades Hospital in Paris. Laennec invented it because he was uncomfortable placing his ear on women's chests to hear heart sounds.
4. 'Cherryes in the ryse'—'Cherries on the bough'.
5. It is interesting to note that Paris, then as subsequently, was seen as the arbiter of sophistication in couture!
6. 'Canwyke'—'Candlewick'.
7. 'ryster grene'—'rushes green'.

4

The Music of Where We Are: How Place Shapes Sound

Songs of the Earth

We save and stylise the signals that matter to us most in terms of our background. Sound belongs to the place of its origin as an event synonymous and analogous with the environment in which it had its genesis. Yet these sounds have the capacity to travel in the imagination and memory, cross boundaries, frontiers, barriers and great distances of time and space, migrating between cultures and taking a landscape with them. This chapter will concern itself almost exclusively with the expression of this as characterised by the music that grows out of a place, because it is primarily through this means of expression that our formative sounds of a home or homeland may be dramatised, romanticised, characterised and transported to other locations. Thus a social and cultural document is created, containing the imaginative construct of a specific place, be it a village, a town, a region, a country or a people, making it eloquently communicable across years and miles, to other cultures. In so doing, it becomes a reminiscence, or an escape and a protest against change and oppression, while preserving images of what may have been left behind, and what would be otherwise lost. It begins in our most intimate experiences of

© The Author(s) 2019
S. Street, *The Sound inside the Silence*, Palgrave Studies in Sound,
https://doi.org/10.1007/978-981-13-8449-3_4

locality and personal environment. Exploring the origins of our sense of place, Michael Stocker notes that 'we…seek acoustic assurance of safety – represented in the sounds of shelter and laughter, of cooking and of song; the pattering of rain on the roof and the crackling fire in the hearth. These comforting sounds – almost as indelible to our species as the sounds of heartbeats or breathing, give us messages of security, grounding, abundance and growth' (Stocker, 21). The first sound world from which we are exiled is the womb, and the first songs we hear when we leave it are lullabies sung by our mother. 'It is through hearing the sound of the mother's voice that you know where you are in the world. It is through this voice that you feel comfort and security. We think of our mother tongue as the language in which your mother first spoke or sang to you' (Seidler in Bull and Back, 399). The very phrase 'mother tongue' is a human connection to a place and its traditions, akin to such expressions as 'homeland' and 'mother earth'. Our first excursions as children show us the immediate vistas that surround us, with their sounds and voices, and as we grow, we remember them.

For the English composer Edward Elgar, music almost literally contained the sound of the homeland earth of the Malvern Hills in Worcestershire, a beloved place all his life. The son of a piano tuner and violinist who ran a music shop in the cathedral city of Worcester, he grew up within sight of this great range of hills that rears up close to the borderlands of England and Wales. The famous film by Ken Russell shows him galloping on horseback across these hills, cycling and flying kites as a boy. There is no question but that the place is in his music: one only has to listen to the vibrant *Introduction and Allegro for Strings* to imagine the winds blowing the grasses as he stood looking out across the landscape, looking for the sun's glint on the River Severn, another of his boyhood haunts. In an adult letter to his friend Sidney Colvin, written in 1921, he said 'I am still at heart the dreamy child who used to be found in the reeds by Severn side with a sheet of paper trying to fix the sounds and longing for something great' (Price, 80). As his career burgeoned, there were many professional enticements to move to London, not least by his wife Caroline Alice, but when we examine his oeuvre it becomes clear how much would have been lost had he done so. His mother Ann, a huge influence on him, had encouraged him to make a study of an Iron Age hill fort on the Malverns, known as the British Camp, or Herefordshire Beacon. It was here that an ancient chieftain, Caractacus, was defeated by the Roman occupying forces and was taken as a prisoner to Rome for trial, where he was eventually pardoned by the Emperor Claudius. We may think of the young Edward walking around the site, conjuring in his mind the sounds of distant lives, and blending his thoughts and fancies with the winds that still blow around the place. This imaginary sound world later inspired Elgar's work, *Caractacus*, a cantata in six scenes begun shortly after the diamond jubilee of Queen Victoria. The third scene opens with an interlude that is full of the sounds of woodland areas around his home. Thus we can say that for Elgar, the inspiration lay almost literally in the soil of the very place. It is hard for the cultural tourist to visit this part of England, without 'hearing' Elgar's music, be it the *Enigma Variations*, the *Cello Concerto* or some of his great choral works, such as *The Dream of Gerontius*, just as to see even a photograph of a famous cave on the Scottish Isle of Staffa will not conjure for many the opening of Mendelssohn's concert overture, *The Hebrides*.

Felix Mendelssohn was just 20 years old when he and his friend Carl Klingemann who was secretary to the Hanoverian Legation in London took a holiday in Scotland, 'with a rake for folksongs, an ear for the lovely and a heart for the bare legs of the natives' (ibid., 17). A visit to Holyrood Palace yielded an idea that was to become the germ of his *Scottish Symphony*, but it was on the 7 August 1830 that the two men took a boat trip to *The Hebrides* that prompted—apart from severe seasickness— Mendelssohn's inspiration to write his most famous and evocative tone poem. On the 8 August, they landed on the uninhabited island of Staffa, with its breathtaking and mysterious cavern known as 'Fingal's Cave', so-called after the eponymous hero of a poem by the eighteenth-century Scottish historian-poet, James Macpherson, who had claimed in 1761 to have 'discovered' an epic text on the subject of an ancient hero, related, so he suggested, to the Irish mythological character, Finn McCool, and according to Macpherson, written by a bard called Ossian. This took the form of *Fingal,* published under that name in December of the same year. The full title was: *Fingal, an Ancient Epic Poem in Six Books*, together with Several Other Poems composed by Ossian, the son of Fingal, translated from the Gaelic Language. Upon publication, the authenticity of these writings was immediately challenged, and although he never denied his claims for their origins, Macpherson failed to ever produce evidence of actual manuscripts; it seems likely therefore that the whole thing was a constructed fiction in the tradition of the English forger poet, Thomas Chatterton. Nevertheless, Macpherson's work, if it is his, echoes with certain music of its own from the very first line: 'Carthullia sat by Tura's wall, by the tree of the rustling sound' (Macpherson, 1). Later Carthullia commands the minstrel, Carril, to sing the story of Fingal:

> His followers were the roar of a thousand streams…He sat in the hall of his shells in Lochlin's woody land…Three days within my halls shall we feast…that your fame may reach the maid who dwells in the secret hall… (ibid., 21–2)

It is easy to understand how this epic, combined with such a spectacular landscape and seascape, might seize the imagination, and the Staffa cave, pillared after all by the same rock formation as Ireland's Giant's Causeway across the water in County Antrim, provided the perfect link between myth and geography. At the time of Mendelssohn's visit, the cave had long been part of the itinerary for many cultural tourists; Jules Verne visited and used it in his book, *Le Rayon Vert (The Green Ray)* and also mentions it in *Journey to the Centre of the Earth* and *The Mysterious Island*. Poets John Keats, William Wordsworth and Alfred, Lord Tennyson made pilgrimages, as did the artist J. M. W. Turner, who painted *Staffa, Fingal's Cave* in 1832. Even Queen Victoria herself made the trip. For Mendelssohn and his friend, the effect of seeing the giant black basalt columns, full of towering symmetry, was seminal, Klingemann likening the overall affect to being 'like the inside of an immense organ, black and resounding, and absolutely without purpose, and quite alone, the wide gray sea within and without' (ibid., 18). For the listener, it is all there in Mendelssohn's music. The place caught the imagination, and the young composer transmits the place to the listening imagination via melody, orchestration, timbre and musical dynamics.

*   *   *

What do we imagine when we make pilgrimages of a place that was made famous through the reflections of its sound in popular culture, be it Liverpool and The Beatles, Manchester and The Smiths or California and The Beach Boys? We may associate Memphis with Rock 'n' Roll, Nashville with Country Music or Vienna with Brahms, Schoenberg or the Strauss family; each terrain has imaginative links with its musical sounds, as both formative and reflective soundscapes. What does the *made* sound contain that evokes the place, and how much has the creative imagination of the composer and/or artist been fed by that environment? The musical conduits that brought popular music from the USA via mercantile shipping through British ports such as Liverpool fuelled the development of phenomena such as rise of Merseybeat in the 1960s and offered expressions of urban black music that found an empathy in the minds of youth, while at the same time creating mental images of their origins, and the poetry contained within their very names—New Orleans, St. Louis, Memphis and many more. The relationship between the creation of music—particularly urban popular music—and its place of origin may often be one of rebellion and protest against situation and circumstance as much as a celebration of its environment of origin. In some cases, it can be both, or more. Black American music found expression in Gospel, the Blues as well as in parody (as in Ragtime music, the sound accompanying the Cakewalk, that derided the affected white dancing of the time).

In Britain, there can be no more eloquent recent expression of a home-grown imaginative sound interpretation of place than a statement made by Rob Halford, the lead singer of the Heavy Metal band, Judas Priest: 'Metal came from foundries where the Midlands sound unfurled' (Halford, quoted by Weinstein in Lashua, Spracklen and Wagg, 39). Halford had grown up in Walsall, a place known not only for iron works, but also for the leather clothing industry. In his stage act, he was noted for wearing leatherwear, studded with metal. This is high-octane, high-volume, guitar-based rock music, with a predominantly male following. The macho image was as much a reflection of prevailing local attitudes as its industrial background, although both came from the same source: 'One of the most important features of industrial Birmingham for heavy metal music was the city's decidedly working-class character. So much of its class culture relates to qualities that are part of heavy metal's sensibility: a sense of solidarity, male camaraderie, masculine toughness, and copious amounts of beer' (Weinstein, ibid., 41). It is hard to conceive of such a music culture coming out of, say, southern California. As Weinstein reminds us, 'geography has a greater meaning than just a place on a map. In twentieth-century England, the divide between the south and the rest of the country – political and financial power in the south, contrasted with mere manpower elsewhere – creates a territorial cultural-class split' (ibid., 43).

Similarly, the Manchester band, The Smiths, notably between the years 1982 and 1987, were products of the political culture of a particular time as well as being firmly located in a place for which responses to establishment politics were intensely relevant. Active approximately during the second term of office of the British Prime Minister Margaret Thatcher and her socially divisive Tory government, The Smiths were also strongly linked to the city in which they had grown up, 'the sons and heirs of a heritage of left-wing creative production in that city and its surrounding townships' (Atkinson in Franklin, Chignell and Skoog, 72). Atkinson underlines the relationship between time and place, politics and cultural expression:

> A strong provincial punk scene in Manchester gave birth to a number of bands with diverse styles, linked largely by their geographical origins in Greater Manchester but also by a dystopian imagery…Self-consciously regarding his own background, the [Smiths'] lyricist Morrissey reconstituted images of Manchester and other aspects of British life, in the creation of songs that critiqued many aspects of contemporary eighties life. (ibid., 86)

The history of a place shapes its own cultural tradition that passes through generations; in the case of Greater Manchester during the 1980s, there were 'continuities between the work of The Smiths and the radical working class theatre produced by Ewan MacColl and Joan Littlewood' decades earlier. 'This Manchester and Salford heritage of political activism and left-wing representation was also a feature of the BBC North Region [based in Manchester] in the 1930s' (ibid., 86–7).

Issues of regionality, ethnicity and gender are crucial to the identity of who we are, and the brand loyalty of place, from sea shanties sung on sailing ships to the indigenous folk music of counties, countries and nations, the identification with nationhood and protest, are parts of a discussion that is at once global and frequently intensely local. As the twentieth century began, there was a revival of interest in the folk song heritage in Britain, with collectors such as Cecil Sharp, Maud Karpeles, George Butterworth and Ralph Vaughan Williams seeking to preserve something of the legacy of an ageing generation who retained in many cases, the last vestiges of a tradition that was intimately linked to place in an extremely localised sense. In doing so, they incorporated the use of ancient musical modes into their own music that gave their orchestral adaptations a completely different sound to the music of composers such as Edward Elgar; it seemed somehow to evince images that themselves originated in a folk memory rooted in place, drawn out of the past by hereditary sharing and oral communication. In this sense, it was the human equivalent of birdsong. Sharp collected songs in the Appalachian Mountains of the USA, while writers such as Thomas Hardy, himself a fiddle player and part of a family tradition of village musicians in Dorset, tapped into folk memory and preserving, in fiction, poetic and documentary form, a unique line of audible reflection linking communities to their immediate environments and society. Before this, there had been collectors such as the poet John Clare (1793–1864) who recorded on paper the music of village life around his native Northamptonshire village of Helpston in manuscripts of the songs themselves, as well as in reflections of their cultural influence in his poetry and prose. Clare was for a time taken up by the fashionable London literati, dubbed 'The Peasant Poet' and ultimately found himself isolated as a curiosity and novelty, as interest in his work faded. He remains one of the great poets of the English lyrical tradition and showed in all his writing an acute and highly knowledgeable awareness of the natural world. In Clare, we can almost literally hear the landscape of early to mid-nineteenth-century England, and in some cases, precise moments of everyday village life, as in an undated manuscript account of an ordinary morning in his village. Clare seldom used punctuation, but the sense of immediacy and place is palpable and moving in his writing. Coming through the window of his Helpston cottage he hears:

> …the whistle of the ploughboy past the window making himself merry and trying to make the dull weather dance to a very pleasant tune which I know well and yet cannot recollect the song but there are hundreds of these pleasant tunes familiar to the plough and the splashing stream and the little fields of spring that have lain out the brown rest of winter and green into mirth with the sprouting grain the songs of the sky lark and the old songs and ballads that ever occupy field happiness in following the plough – [but] neither heard known or noticed by all the world beside. (Quoted by Deacon, 74)

It is a vivid snapshot that in turn evokes a considered reflection by Clare on the nature of the extreme localness of rural society before the transport revolution that would change social mobility and the character of communities from the mid-nineteenth century onwards. Perhaps the most significant part of this lovely passage is the very last line: '…neither heard known or noticed by all the world beside'.

Mystery Train

Where we are is much more than physical: culture, background and circumstance all find outlets in sonic expression. Although slavery was officially abolished in 1807, an illicit trade in the USA continued until the American Civil War of 1861–1867. Some fifty years earlier, in 1816, a witness George Pinckard described hearing the songs of the inhabitants of a recently arrived ship:

> They have great amusement in collecting together in groups and singing their favourite African songs; the energy of their action is more remarkable than the harmony of their music…We saw them dance and heard them sing. In dancing, they scarcely move their feet, but threw about their arms, and twisted and writhed their bodies into a multitude of disgusting and indecent attitudes. Their song was a wild yell devoid of all softness and harmony, and loudly chanted in harsh harmony.' (Pinckard, quoted in Oakley, 14–5)

Torn out of their environment, bewildered and with virtually nothing in terms of possessions, it is little wonder that these people clung to each other and the recollection of tradition. Yet the white slave owners soon put an end to such tribal and communal activities. Tribes and related groups were split up and 'the Black Codes of Mississippi…put an end to the beating of drums out of fear that the slaves could communicate and concert a revolt' (Oakley, 15). Work songs would become the only licensed form of music making, but it did not take long before these began to evolve into a new hybrid music, as was noted by Nicholas Cresswell in his *Journal* 1774–1777, in which he 'described slaves in Maryland dancing to a banjo made out of a gourd "something in the imitation of a Guitar, with only four strings" (Quoted by Oakley, 15). They would sing in a "very satirical manner" about the way they were treated. "Their poetry is like the Music – Rude and uncultivated"' (ibid.). Through such accounts do we hear the beginnings of what became known as the Blues and Gospel. In Christianity, slaves found a specific personality, a victim of persecution 'who had suffered as they suffered…They so transformed the religion of the slave owner that eventually they came to look down upon the white preachers and white religious services…and through the religious songs they made up from Biblical stories, they expressed their real feelings about slavery' (Lester, 79/83). Examples of what became known as 'Negro Spirituals' are numerous and have passed into the popular mainstream:

> Joshua fit the battle of Jericho,
> Jericho, Jericho.
> Joshua fit the battle of Jericho,
> And the walls come tumbling down. (Lester, 83–4)

Some such songs carried double meanings; for example, the spiritual, 'Steal Away', was, read on one level, as a promise of salvation, after premature death:

> Steal away, steal away, steal away to Jesus.
> Steal away, steal away home
> I ain't got long to stay here…(ibid., 105)

Yet, as an anonymous account held in the US Library of Congress, and quoted by Julius Lester suggests, it also contained a code, 'that mean there going to be a religious meeting that night. The masters before and after freedom didn't like them religious meetings, so us naturally slips off at night, down in the bottoms or somewheres. Sometimes us sing and pray all night' (ibid.).

Gospel music grew out of these sometimes raucously interactive services, a release from the constrictive bondage in which they spent most of their working lives, and the power of black choirs as they evolved contained in their assertiveness a triumphant belief, hope and defiance expressed through their very vibrancy. The other side of the coin was a solitary one in the form of Blues music, an expression of despair, and escape through alcohol, or sex, and the retribution and betrayal that came from both. It was sometimes music that combined with religious searching, as heard today through recordings by such singers as the Reverend Gary Davis. Born in 1896, Davis had begun as a Blues singer but converted to Christianity in the 1930s, becoming ordained as a Baptist minister in 1937. Yet even brief historical snapshots such as these can only give a slight taste of what evolved from a social situation as much as a single style. In the words of Francis Davis, 'nothing is that simple…Let's just say the Blues has a large and extended family and bears a strong resemblance to many of its relatives. Especially in its younger years, it was frequently sighted in places it may or may not have been, and mistaken for blood relatives long since forgotten' (Davis, 57). Early recordings show us the beginnings of the blending of work song, Gospel, Blues into jazz and beyond that Rhythm and Blues and Rock 'n' Roll. Above all it is folk music and reflects the time in the places that shaped it. It is a response to social conditions and a human need to express that response, and it lies, of course, in a shadowy aural world of memory and legacy, long before the advent of recordings.

American popular music has continued to express its time and find voices of either escape or a shared attitude; whether it be in the music of prohibition, the Depression, war or protest. The coming of the mass commercial reproduction of sounds and the marketing of such music ultimately to a global audience sold it to audiences with no direct connection to its history and origins, and in many cases, evolved it away from its roots. Jazz music in its earliest incarnations existed on the fringes of popular culture, consumed mostly by black communities, and its origins owed much to the inborn ability of its performers to improvise, a skill learned and honed under oppression. Although claimed by many to be 'America's music', as we have seen, once we link its creation to the Blues and beyond, we find ourselves locating it in another imaginative world altogether. Like the Blues, pure jazz, from its early evolution in New Orleans during the 1890s and 1900s, has usually been a niche musical taste, although its hybrids have expressed themselves increasingly through mass hybrid forms. 'With the advent of swing in the 1930s, jazz came the closest it has ever come to widespread popularity, that brief period called the Swing Era, roughly 1935 to the early 1940s. But swing itself…exists as an amalgam, a mix of jazz, dance music, popular songs, and standards that receive a rhythmic emphasis that causes them to "swing"' (Young and Young, 118). Here the music came at a time when social requirements were governed by an age of anxiety; it had become part of a global relief from Depression and the threats and privations of war, was disseminated by recordings, radio and film and while belonging to no place in particular, yet espoused ideas, images and cultures that replaced hardship with illusory glamour and wealth.

Development of 'the people's music' is as much about appropriation as innovation; quoting an interview with Elvis Presley, the writer Tony Russell highlights the influence of black music in Presley's singing, particularly in the early recordings made for the Sun Record label in Memphis, Tennessee:

> We were a religious family…going around together to sing at camp meetings and revivals. Since I was two years old, all I knew was Gospel music, that was music to me. We borrowed the style of our psalm singing from the early Negroes. We used to go to these religious singings all the time. The preachers cut up all over the place, jumping on the piano, moving every which way. The audience liked them. I guess I learned from them. I loved the music. It became such a part of my life it was as natural as dancing, a way to escape from the problems and my way of release. (Presley, quoted by Russell, 161)

Such cultural 'borrowing' was by no means new; Dvorak in the nineteenth century used black music as themes in his *Symphony No. 9*, subtitled *From the New World*, and prior to the outbreak of the First World War, in Britain Ralph Vaughan Williams, Gustav Holst, Frederick Delius and others, as we have seen, used their folk song collecting to enhance new works, adapting ancient folk airs for the concert hall. In so doing, the identity of localness is disseminated, while being changed and diluted, and the places of origin become in some cases globally famous, and destinations of pilgrimage for cultural tourists hungry to touch the taproot and drink from the authentic source. Yet in so doing, the superficial searcher after the inspiration of the place will often encounter a construct, a kind of theatre that preserves an illusion of a source that is no longer there, or in some cases, never was. The global audience for the Viennese New Years Day concerts, and those wealthy enough to possess a ticket to sit in the glittering, golden Musikverein, are served a confection of waltzes by the Strauss family. Yet there are marches too; the culture that created this sonic opulence was a military one, and the very buildings that house the music are relics of an empire built on power, war and sometimes racism, within expressions of nationalism.

*   *   *

Equally, the musical heritage celebrated in modern Memphis, USA, to which the tourists flock from all over the world, is largely a manufactured experience; Beale Street, Graceland, Sam Phillips' Sun recording studio and the headquarters of Stax Records are marketed as the spiritual base that bred what the guide books call 'America's music'. Yet appearances can be deceptive:

> In 1977, when the first major wave of visitors arrived in Memphis after the death of Elvis Presley, there was little to see…Beale Street and downtown had become a ghost town through demolition created by urban renewal, which city leaders deliberately accelerated following the 1968 assassination of Dr Martin Luther King Jr. Graceland, where Elvis spent the last 20 years of his life, and where he died, was inoperable. The building on Union Avenue that housed Sun Studio was unrecognisable, and Stax Records, closed by court order in 1976 bankruptcy proceedings, awaited sale to its new owner, the Southside Church of God in Christ, which purchased the property in

1980 and demolished the building in 1988. (Rushing in Lashua, Spracklen and Wagg, 259–60)

Jim Jarmusch's film *Mystery Train* provides a flavour of Memphis before the makeover. In order to preserve a physical presence to match the sonic imagination, all over the world, where there is financial reward in feeding an appetite for the inherited sound of a place, local authorities create stage sets upon which are played out dramatised reminiscences approximating, in various degrees, to the original generative creative impulses of the locale. As far as Memphis is concerned, 'today, city leaders and developers, who in earlier times might have distanced themselves from local culture, now boast about the global significance of the culture and the place associated with it' (ibid., 261).

The Beat Goes On

As part of its ongoing *Creative Cities Network*, UNESCO named Liverpool as a 'City of Music' in 2015, joining a select although evolving group of other cities around the world to jointly hold the title. It is instructive to note the others in the list:

Almaty (Kazakhstan)
Amarante (Portugal)
Aukland (New Zealand)
Brno (Czechia)
Chennai (India)
Daegu Metropolitan City (Republic of Korea)
Frutillar (Chile)
Glasgow (Scotland)
Kansas City (USA)
Morelia (Mexico)
Norrköping (Sweden)
Pesaro (Italy) and
Praia (Cabo Verde).

The initiative behind the Creative Cities Network, at the time of writing numbering 180 locations, includes various designated areas of creative interest other than music, such as design and media. It was created in 2004 'to promote cooperation with and among cities that have identified creativity as a strategic factor for sustainable urban development. The…cities that currently make up this network work together towards a common objective: placing creativity and cultural industries at the heart of their development plans at the local level and cooperating actively at the international level'.[1] Thus the migration of localised expression continues to blend, some might say, at the risk of dilution, while others would point to the strong identities of the cities chosen, and the individual sound that remains a part of their physical and cultural beings. We move through the world as we hear it; we cannot escape being in a place, and the sound of the place reverberates in our consciousness, finding expression through attitude, temperament and our own evolving characteristics as human beings. The sound of where we are and where we have been, the formative physical environments that surround us, govern the sounds we make, and the sounds with which we empathise. As Fran Tonkiss has written: 'Sound gives us the city as matter and as memory. In this register, the double life of cities – the way they slide between the material and the perceptual, the hard and the soft – is spoken out loud, made audible' (Tonkiss in Bull and Back, 303). It is defined in one way as 'the clamour, the density, the sheer weight' of it. 'And then – listen – there is the way a city comes to us in memory and reverie, its cadences, whispers [and] sighs' (ibid.).

The music business has always been quick to create brands around such triggers, as with the music that emerged from Liverpool and Detroit in the 1960s. Popular music is based on oral and *aural* history, and what we hear shapes what we make, absorbing from the past and melding tradition with environmental sounds, both natural and manufactured, intentional and incidental. In the Southern States of the USA, Country Music met Blues and Gospel, through a filter of urban sounds familiar in particular to young people in deprived or poor areas. Listen to the train imitations and rhythms in the music of Sonny Terry and Brownie McGhee, or Chuck Berry for example. As Tony Russell points out:

A musician would be open to sounds from every direction: from family and friends, from field and railroad yard, lumber camp and mine; from street singers and travelling-show musicians; from phonograph records and radio; from dances and suppers and camp meetings and carnivals; from fellow prisoners in jails, from fellow workers everywhere. (Russell, 10)

The 'Motown Sound' created by Lamont Dozier and Eddie and Brian Holland famously 'developed a rhythm based on the clattering mechanical beat of Detroit's automobile assembly lines (where many of the Motown staff had worked,) and created rudimentary sound effects with chains, hammers, and planks of wood, or by stomping on floorboards' (Perry and Glinert, 191). In turn, travelling merchant seamen returning home to Liverpool imported this and other music they had heard while in the USA, where it was assimilated by music-hungry teenagers, growing up in a city with its own strong indigenous musical influences shaped by immigration, in particular from Ireland and Wales. The cross-fertilisation of cultures through travel as it became more global made a new 'local' sound that was to become itself universal, influential and of course marketable. It is interesting to note that, even now, in Liverpool, The Beatles are considered to be a 'local' band, and the burgeoning of their success, as in a way a departure from their roots. Influence and circumstance, in partnership with invention, shape the sounds of a place; urban music grows where it is seeded and 'cities provide a soundstage for the drama of modern life' (Tonkiss in Bull and Back, 304).

In England, this sense of cultural identity is particularly strong in the North, where individual city cultures are diverse, unique and fiercely guarded and defended. At the same time, 'music in the North of England is appreciated for its "placeness" [while] on the other hand it is renowned for its openness to foreign influences. Its cosmopolitan character in part results from its history and in part from its desire to resist the pressure to conform to the values and fashions coming from the South' (Mazierska in Mazierska, 5). Like Liverpool, Manchester is a case in point:

Manchester is a hybrid town, born all in a rush one hundred and fifty years ago, when those looking for work in the fast-growing factories, workshops, warehouses and foundries included large numbers of Catholic Irish, as well as Scots, and German and East European Jews. These migrations have been replicated since, with incomers from the Caribbean in the 1950s and from the Asian sub-continent in the 1950s. (Haslam, xi)

Such touching-places occur throughout the world; southern Spain, for example, has a culture influenced by ancient influence, invasion, immigration and mercantile imperatives from around the basin of the Mediterranean, and notably from Africa. The sound is part of the history and the culture, and the music grows out of the sound. In a previous chapter, we reflected on the power of bells within the historic identity of a place. Today, we may struggle to hear those sounds, and indeed human voices as Ben Jonson did in Elizabethan London, above the noise of the modern world, as Tonkiss reminds us, the *musique concrète* of place constantly evolves:

> The special sense of a city may be no longer given by tower-clocks and church-bells – by sounds, that is, that tell the time – but rather by those that tell of motion. The peculiar sounds of transit are the signature tunes of modern cities. These are sounds that remind us the city is a sort of machine. The diesel stammer of London taxis, the wheeze of its buses. The clatter of the Melbourne tram. The two-stroke sputter of Rome. The note that sounds as the doors shut on the Paris metro, and the flick, flick, flick of the handles. The many sirens of different cities. (Tonkiss in Bull and Back, 306)

Even in these sounds, and the music that comes from them, a place retains and transmits its character to the imagination, and an unsighted person, familiar with international travel, could well identify their location by the ambience around them. We await the next revolution, as with the coming of electric transport, our cities grow quieter again; will they then be defined once more by bells and voices? And how will their music sound then?

Making the Place Sing

An apparently near-silent space may evoke sound. There is the desire even in the youngest child to 'test' the acoustic response of a great interior, to hear themselves amplified, echoed and enhanced by their surroundings. Gothic churches and other sacred spaces are transformative sound studios, and a composer writing music for these giant chambers must hear imaginatively even before committing a note to paper, the interaction of the stone with the timings, assonances and forces involved in the interaction. Certain spaces affect the sonic imagination, as certain styles of architecture engender spiritual inspiration within resonances, their roles as architectural 'voices' in the performative sense, a combination of sung liturgy and architectural reverberation that creates an aural experience heard as both transcendent and mystical. Throughout history, we have been at great lengths to create places of refuge in which to contemplate, or to listen to sounds that belong nowhere else.

Today, much effort is expended by acoustics technicians in making sonically specialised environments in which to best absorb music and/or language. Of these, our churches have always been machines of sound, at one and the same time, resonating boards and sonic instruments in their own right. It is salutary to recognise the length and depth of the tradition to which this preoccupation belongs.

The relationship between who we are and where we are is—and always has been—linked intimately to the very structures we build for our habitation, entertainment and worship, the contexts for our day-to-day living. The Roman author and architect, Marcus Vitruvius Pollio, commonly known as Vitruvius, who lived from approximately -80BC or -70BC to around -15BC, provided in his writings—particularly in *The Ten Books on Architecture*—detailed explanations of the fundamental connections between stone and sound, and acknowledge their origination to be from even earlier times, linking dimensions, ratios and proportions to the human form itself:

> …Since nature has designed the human body so that its members are duly proportioned to the frame as a whole, it appears that the ancients had good reason for their rule, that in perfect buildings the different members must be in exact symmetrical relations to the whole general scheme. Hence, while transmitting to us the proper arrangements for buildings of all kinds, they were particularly careful to do so in the case of temples of the gods, buildings in which merits and faults usually last forever. (Vitruvius, 73)

Vitruvius also visualised sounding aids in other public spaces, such as theatres, by means of which:

> ...the voice, uttered from the stage as from a centre, and spreading and striking against the cavities of the different vessels, as it comes in contact with them, will be increased in clearness of sound, and will wake an harmonious note in unison with itself. (ibid.,143)

Much later, stately homes across Europe would be built with minstrel galleries, where musicians might play to accompany banquets and gatherings, an early form of ambient music that might or might not be designed to be listened to as opposed to being simply heard as background. The English artist and architect John Shute (died 1563) showed his debt to Vitruvius in his book, *The First and Chief Grounds of Architecture*, published in the year of his death:

> Musicke...is verie necessary for an Architecte, for these causes must have, as it were a foresight in it, that therby the principall chambers of the house, shuld with suche oder be made, that the voice or noyse of musicall Instrumentes, should have their perfaict Echo, resounding pleasauntly to the eares of those that shalbe hearers therof, as also the Romaines, used in all their palaces and for many other necessities therunto belonging, of the which Vitruvius, maketh further demonstration, as the refreshing of the Melancholicke mindes, which are always travailing for further knowlaige. (Shute, 34)

The German Jesuit scholar and polymath Athanasius Kircher (1602–1680) explored the world of acoustics in great and often fanciful detail, creating complex architectural designs in which buildings literally had ears (and voices) through tunnels, resonators and tubes along which sound and echoes could be transmitted from room to room. Of particular interest to Kircher was the sonic potential of spiral tubes as sound conductors through masonry, as explained in his *Musurgia Universalis*, written in 1650:

> One may well wonder why the multiplication of sound is so strong in a cone twisted in a spiral. I have certainly pondered long over this matter,

finding at last that a helical cone, twisted in a certain ratio, makes some kind of parabola which brings about infinite conglomerations of sound. So it is no wonder that it achieves such energy in the multiplication of sound. (Kircher, quoted in Godwin, 165)

For Kircher, such elaborate structures were the very mechanics of the sonic imagination; he described many of his experiments in his *Phonurgia Nova* of 1673, as Godwin explains:

Kircher…conducted acoustical experiments at the sanctuary of Mentorella, which stands on a spur overlooking a wide valley dotted with villages [near Rome.] He found that by using a "cleverly constructed" tube…a voice could be heard at Toretta, two miles away; Siciliano (now Ciciliano), three miles…and even Giramo (Gerano) five miles. (Godwin, 168)

Thousands hearing the sounds emitting from the holy place responded and obeyed the commands of what they believed to be the oracle of the shrine, while 'the local wolves, too, answered when one of his friends tried howling through the loudspeaker' (ibid.). We shall explore the phenomena of unexplained and apparently supernatural sound further in Chapter 6 of this book. Suffice to say here that place and sound, from landscape down to the actual fabric of buildings, have always been connected with ourselves as contextual environments, in which certain sonic rules and conformities are expected, and within which therefore, the unfamiliar can chill us with its alien nature. Likewise, the apparent freedom of speech within one's own environment, if undermined, may be the root of paranoia. Through turbulent eras of history, devices such as Kircher's have been adapted to enable clandestine listening through walls. The very term 'eavesdropping' has its origins in devices known as 'eavesdrops' built into the eaves of such buildings as Hampton Court by Henry VIII to demonstrate that every statement had the potential for being overheard. Walls truly did have ears, and Place and Person might share a mutual paranoia and suspicion.

Unsurprisingly it is in ecclesiastical buildings that sound and imagination have always met with the greatest potency. In July 1845, the English cleric William Gresley preached a sermon at the consecration of St Mark's Church, Great Wyrley, in Staffordshire. St Mark's is a small, unostentatious building, but for Gresley, it was much more: 'What was yesterday but a heap of stones, is now God's House – separated from common uses'. Warming to his theme, he continued, making the point strongly that a church '...ought to be a building *distinct in character* from all others – distinct from our private dwelling houses – our school-rooms, our theatres, our lecture rooms. It should at once convey the impression to he who enters it that he is on holy ground' (quoted by Whyte, 65/68). Although he did not mention it, Gresley might well have identified acoustics as one of the elements by which sacred buildings announce themselves. Many such buildings around the world have been constructed with the specific imaginative intention of creating often deliberately spectacular sonic effects, just as their art has sometimes told graphic and salutary stories in order to subjugate and elevate their faithful visitors. The Venetian churches of Palladio and others were designed with minute attention to their sound characteristics, and the music composed to be played and sung in them cannot communicate in the same way when heard in spaces where these sonorities are not considered. In Lucy M. Boston's haunting novel, *The Children of Green Knowe*, there is a powerful example of how the space contained within the walls of a great cathedral impels a response. The book revolves around the lives of four children, one from the present and three from Restoration England. Here, one of the children, Alexander, alone in a great English cathedral, finds himself overcome with the desire to respond to the physical space around him:

> He strayed from his family the better to concentrate on the sensation of tingling emptiness and expectation in the building that he found so strange and so enthralling...The candles waved in the air that was as much in movement as if in a forest. Every now and then a spindle of wax breaking off a guttering candle fell into the brass holder with a bell-like note that seemed to go up and up and be received into Heaven. Alexander held his breath and listened. There was no sound except a low droning of wind passing along the distant vaulting, the kind of sound that is in a shell. He had a sudden great desire to sing, to send his voice away up there and hear what nestling echoes it would brush off the roof, how it would be rounded and coloured as it came back...He tossed the notes up...He could feel the building round him alive and trembling with sound...It was as if the notes went up like rocket stars, hovered a second and burst into sparklets. The

shivered echo multiplied itself by thousands. One would have thought every stone in the building stirred and murmured. (Boston 2013, 131–3)

It was Goethe who wrote, 'I call architecture "petrified music"' (Howard and Moretti, 8). Alexander effectively has a mystical experience, but it has been a man-made one; he is responding to the intentions built into the place by architects and artisans, a *manufactured* studio designed to achieve precisely such an effect. There are sacred places such as Istanbul's great mosque, the Hagia Sophia, likewise shaped for such purposes (among others). A Greek Orthodox Cathedral between 537 and 1453 AD, prior to Constantinople being conquered by the Ottoman Empire, when it became a mosque, the Hagia is now a museum. Although today no musical performances are allowed there, researchers from Stanford University have created a digital experiment in order to simulate what worship would have sounded like when it was a medieval cathedral, by recording a virtual performance elsewhere, while synthesising the acoustics of the original building. By doing so, not only have they post-processed the sound to give their singers the sound of being in the Hagia Sophia, they have been able to provide them with real-time audio feedback to enable them to sing as if they were actually in the building. Collaborating with choral group Cappella Romana, the sound specialists digitally recreated the former holy building's acoustics, in the university's Bing Concert Hall, as if it was Hagia Sophia. The project focused on the interior of The Hagia, using recordings of balloon 'pops' taken in the space, and other audio and visual research, to ascertain the building's acoustics by extrapolating from those noises. The scientists used that data to give an impression of the experience of being there. To recreate the unique sound, performers sang while listening to the simulated acoustics of Hagia Sophia through earphones. Their singing was then put through the same acoustic processor and played during the live performance through speakers in the concert hall, as they also sang, giving the performance the effect of it taking place in the Hagia Sophia itself, an illusion which involved slowing of tempo in order to adjust to an eleven-second reverberation time. Thus current technology created an illusion of a sound unique to a space like no other on earth, in a recording that took place thousands of miles from the place itself. It would seem that an experience of almost miraculous transcendence CAN be replicated, and 'petrified music' be released once more into time. Yet added to the sophisticated acoustical aids supplied by the scientists, undoubtedly the singers themselves would have had to imaginatively transport themselves into the ecclesiastical space, mentally picturing the architecture, and placing their performance within a simulated location present in their mind's eye. To fully achieve such a recreation, the effort need must be psychological, artistic and even spiritual, as well as technical.

In Venice as in Istanbul/Constantinople, church architects building the great churches of the time, such as Palladio Venice and Sansovino, working with composers, could manufacture sonic effects with the most minute attention to detail, in the same way as a modern audio studio recordist or designer would establish a specific sound acoustic to fit the requirements of particular forms of music. More than this, to the tuned ear of the time, these spaces carried their own auditory signature, so for a listener acquainted with the characteristics of a certain building, a few notes from an appropriate piece, say Striggio's mass, *Ecce Beatam Lucem* would have been sufficient to identify the space in which it was being performed. Polyphonic choral music was made for great reverberant spaces, such as the Sistine Chapel, or the great Gothic cathedrals of France. Churchgoers like the young Alexander in Lucy M. Boston's *Children of Green Knowe* would have absorbed the song of the place in which they found themselves, and convinced themselves that it must be sacred, because although it was uttered through human agency, it was transformed by the very walls of 'God's House' itself. Here could be heard the illusion of a kind of music of the spheres, but miraculously coming to them through their own ears, even their own voices transformed in a physical manifestation of sound through which sacred air became suddenly audible.

> *I call, I call, I call,* (he sang)
> *Gabriel! Gabriel! Gabriel!*
> …He tried it again, louder.
> *Gabriel! Gabriel! Gabriel!*
> He could almost imagine the Archangel must hear, might come.
> (Boston, 133)

Alexander is interacting with the edifice. In today's American black Baptist Gospel services, a congregation whose ancestors had been slaves will interact spontaneously with what they hear from the preacher, and make their own voices audible in cries of affirmation and establishing an audible presence for themselves within the framework of what is, in every sense, a religious experience. By being a voice, we become a presence and make ourselves part of the heard moment. Participation with a place, a partnership inspired by imagination, is an act in which we share our own imagination and become part of the imagination of others, while shaping the memory of *now*. Whether it be to assert our selves or to escape our predicament, the sonic circumstances of the world make us who we are, just as we make the sounds that surround us, shaping expressions of being as dramatisations created through imagination, influencing others through the ability to stimulate their sonic imaginations through our own. We are always *somewhere*, always subject to events of sound that have the capacity to change or underline our being. The song of where we are may confirm who we are; equally, it may also remind us of where we have come from, or who we would wish to *be*.

Note

1. UNESCO Creative Cities Network mission statement, https://en.unesco.org/creative-cities/home. Accessed January 2019.

5

Through Storm and Stone: Radio, Sound and the Imagined World

Transmitting Illusions

In the two chapters that follow, we will explore the effect on the imagination of sounds as they are presented within certain media and examine some changing perceptions of alternative realities since the birth of electronic communication. It is notably in radio and its kindred media that poetry and sound have their most intimate kinship. There is a poetic style of making that lends itself in particular to the radio feature. The suggestive nature of the audience makes it a willing accomplice in what may be described as a form of jointly agreed deception, the willing acceptance that lies at the heart of all fictive expression. Some early radio sports commentaries were accompanied by charts of the playing area in listings magazines, to enable the audience to 'picture' the progress of the action through numbered squares. Very soon it became clear that this was not necessary, that the mind could supply pictures that enabled involvement. Today, we may listen to a commentary from another room and gain a sense of action simply by the tone, inflection and sense of excitement conveyed within the commentator's voice. In terms of fiction, sounds imagined on the page move forward to the ear on radio, generating pictures dependent

© The Author(s) 2019
S. Street, *The Sound inside the Silence*, Palgrave Studies in Sound,
https://doi.org/10.1007/978-981-13-8449-3_5

on the skill and subtlety of the practitioner. Radio is indeed a suggestive medium, and it is a failing of less experienced producers to underestimate the capacity of the listener to be a working and creative partner in the process.

We bring words into meaning within our heads both silently and noisily at the same time, and this takes us back to the birth of poetry, and its original orality, as in the singing of Homer and the performance of bardic ballads as performed by troubadours. Radio sound reclaims the oral, without the distractions of the visual. In 1938, the famous CBS Orson Welles War of the Worlds broadcast created the stir it did because the imagination of the audience took over and explored the evidence of their own immediate experience within the context of the 'seeds' of sound that presented the suggestions of reality; thus a listener looking from their window saw an empty street and feared the end of the world, while another might see a crowded street and believe that mass evacuation was underway. The medium of sound can be argued to be the best place in which to suspend disbelief and invoke the imagination; Hamlet says 'We'll hear a play tomorrow' (Act 2, Scene 2). The very word 'audience' derives from the Latin source of 'audire', meaning 'to hear', and Shakespeare himself, understanding the shortcomings of the 'unworthy scaffold' (*Henry V*, prologue) in which he worked as vehicles for spectacle, asked the permission of his audience for a tacit agreement that allowed his actors to 'on your imaginary forces work' (ibid.):

> Think, when we talk of horses, that you see them
> Printing their proud hoofs i' the receiving earth;
> For 'tis your thoughts that now must deck our kings,
> Carry them here and there; jumping o'er times,
> Turning th' accomplishment of many years
> Into an hour-glass… (ibid.)

What Elizabethan theatres could not provide in terms of spectacle, their writers sought to offer in sound through the richness of language, and the physical structures aided the audibility; the very 'O' shape of buildings such as the Rose, the Globe and other London theatres attests to the closeness of the audience to the sound, and its reverberation around the walls. The circular nature of these interiors of wood and plaster 'offered not only visual interest but the resonators and baffles required for good sound distribution in a large space' (Smith, 207). Just as the Chorus in *Henry V* tells us to imagine sound, so the Messenger in the 1623 Folio version of *Coriolanus* shows us that the audience at the Globe premiere of the play in 1608 were asked to imagine much more than they actually heard, given the indication of the stage direction:

Messenger

Why harke you:
Trumpets, Hoboyes, Drums beate, altogether.
The trumpets, sack-buts, Psalteries, and Fifes,
Tabors, and Symboles, and the showting Romans
Make the Sunne dance. (1623 Folio, Act 5, Scene 4, lines 49–52,
quoted in Smith, 242)

As Smith underlines, 'what the audience in fact hears are trumpets, haut-boys, and drums – loud enough in themselves. What the Messenger *tells* them they are hearing is a much wider range of instruments and a volume of sound that, figuratively at least, pushes beyond the theatre's walls to the limits of the cosmos' (Smith, 243). Stage directions are indicative; the realities of performance are governed often by limitations in the physical world; Shakespeare might even have suggested his audience close their eyes, although as far as we know he did not employ quite such a radical suggestion. Yet, when he declares that ''tis your thoughts that now must deck our kings', he is showing us how he would have completely understood the concept of audio drama. In the limitless space of the radio stage, we hear the effect of the dramatist's thought directly transmitted to our brain. The first word on the script of Dylan Thomas's *Under Milk Wood* is an unspoken one: 'Silence'. In the 1952 radio production by Douglas Cleverdon, there then follows about seven seconds of stillness before the first words of the First Voice. It is seven seconds, because Cleverdon judged that to be right, but it remains open to interpretation. However long, we are occupying a darkened theatrical space, awaiting the first spark to ignite the mind, to break into this moment of infinite possibility when the imagination is at its most complicit with the potentials of invention and vision.

Not of the Earth

In its earliest manifestations, 'wireless' transmissions were greeted with a sense of awe by maker and audience alike. On Christmas Eve, 1906, the Canadian inventor Reginald Fessenden conducted an experiment from a station at Brant Rock, Massachusetts—the transmission of speech and music, including a reading from the Bible, a recording of Handel's Largo, and a violin solo. At sea, ships placed at strategic points heard the 'broadcast' and reported back. As it so happened, other wireless operators tuned in by accident, and heard, through the starlit stillness of the Atlantic, voices raised in song and prayer, praising God. It was not too great an exaggeration to say that some believed the source was supernatural; the timing, the content and the apparent technical 'miracle' of the whole thing were overwhelming. Wireless had been used purely for utility until this point; transmissions used Morse code, not words and musical sounds. How could this be anything but some sort of divine intervention? Eighteen years later, with radio established as a mass medium, John Reith, the British Broadcasting Company's Manager and the British Broadcasting Corporation's first Director General, could still write in reverential terms of the power and responsibility of connecting with an invisible force of energy. There was something mystical in this sacred cause:

> When we attempt to deal with ether we are immediately involved in the twilight shades of the borderland; darkness presses in on all sides, and the intensity of the darkness is increased by the illuminations which here and there are shed, as the investigators, candle in hand and advancing step by step, peer into the illimitable unknown. (Reith, 223)

The sense of a medium of the imagination, the inspiration of transmittable thought, permeates Reith's book, as it does many of the other early works written in the white heat of excitement surrounding the birth of broadcasting, others being by two of Reith's senior colleagues at the BBC, C. A. Lewis and Arthur Burrows. Lewis, Reith's organiser of programmes, also wrote with awe of the concept of invisible symptoms of human existence transmitted through space, and in that idea, suggested the potential for the medium to expand literally without limitation:

> The human voice has annihilated space. It becomes endowed with the infinite range of light, for the concerts of last year are still spreading out and on beyond the range of the visible stars. Wireless waves, like light waves, travel at an incredible speed and have greater penetration. Thus, when we speak, it is not to the listener, or even to the world, but to the universe. (Lewis, 144–5)

The BBC's first director of programmes Arthur Burrows began his book with several pages of musings on the image of the universe, blending science and metaphysics, implying that the coming of this new medium might be the very dawn of a new age of discovery: 'The science of wireless and broadcasting…deals with media, agencies, and intervals of space and time which are outside our powers of appreciation, yet have been proved to exist, or have been measured to the satisfaction of many eminent scientists' (Burrows, 3).

* * *

The German Rudolf Arnheim also captured the idealism and essence behind the idea of the new medium in his book, *Radio*, published in 1936 when he wrote:

> This is the great miracle of wireless. The omnipresence of what people are singing or saying anywhere, the overleaping of frontiers, the conquest of spatial isolation, the importation of culture on the waves of ether, the same fare for all, sound in silence. (Arnheim, 14)

The community of air that is evoked in Arnheim's words remains unique. The sense that other people are listening at the moment of hearing is one of radio's many gifts to the imagination, because although we listen, one-to-one, to the message and the voice, yet we are subtly aware that we are part of an audience. There was a time, before radios occupied almost every home, when communal listening in smaller communities was not unusual. Groups of listeners would gather in village halls and community centres, and sit, as at a theatre, to hear live programmes of key events. An amusing newspaper cartoon from those early days shows a man speaking to a woman sitting in front of him and asking her: 'Madam, would you mind removing your hat? I can't hear properly'. That apart, it is still the sense of being alone but part of an imaginary 'family' that is key to overcoming that sense of isolation of which Arnheim speaks. We might argue that with the decline in listening to transmitted radio, and the rise in earpod consumption of downloads and podcasts, we are voluntarily turning our backs on that community. Arnheim understood the poetry inherent in the medium of radio; indeed the sense of transience, of the moment, and the fading to silence as the illusive signal evaporates, that is a part of the poetry, so he is clear that 'poets should emphatically be brought into the wireless studio, for it is much more conceivable that they should be able to adapt a verbal work of art to the limits of the world of space, sound and music' (ibid., 208). He is known as a theorist and perceptual psychologist, and yet, in his description of the radio production environment, he gives us a kind of prose poem of his own:

> The carpeted rooms where no footstep sounds and whose walls deaden the voice…the mystifying ceremonial of the actors in their shirt-sleeves who, as if attracted and repelled by the microphone, alternately approach and withdraw from the surgical charms of the metal stands; whose performance can be watched through a pane of glass far away as in an aquarium, while their voices come strange and near from the control-loudspeaker in the listening room; the serious young man at the control board who with his black knobs turns voices and sounds off and on like a stream of water; the loneliness of the studio where you sit alone with your voice and a scrap of paper and yet before the largest audience that a speaker has ever addressed; the tenderness that affects one for the little dead box suspended from garter-elastic from a ring, richer in treasure and mystery than Portia's three caskets…' (ibid., 19–20)

For some, like Arnheim, the radio studio has always been a kind of temple in which to practise sacred and mysterious rites. For others, it is no more than a workshop. Two producers working in the same organisation may have very different views as to its role and function. In his 1934 book, *The Stuff of Radio*, the BBC producer Lance Sieveking attacked his drama colleague, Val Gielgud for describing the studio dramatic control panel as "'simply the centralising and mixing unit, by means of which the output of several studios can simultaneously be used, and welded together into a single whole at a central point under the control of the producer'" (Gielgud, quoted by Sieveking, 58). Sieveking was working at a time when producers used multi-studio production techniques, in a direct hands-on mode that gave them direct access to the creation of sound expression, without the intermediary of a studio manager. For him, attitudes were contained in the interpretation of terminology, and in his eyes and ears, the control panel was an altar, a conduit for the manipulation of the imagination. 'This is one of *the* essential differences between us. He [Gielgud] thinks the instrument should be "operated," I that it should be "played"' (ibid.).

The poetic imagination that invested radio with wonder for many of those during the development of its possibilities was probably present—and inherent in—the very circumstances of its transmission, and in the reception of the first wireless signals of all, listened for on lonely hilltops through storms, faint signals that defied stone walls in the tiny dots and dashes of Morse code. At times, hard logic seemed to be defied by the laws of nature, and by imaginative possibilities. In the beginning of any new scientific development, very few of its pioneers know the exact limits of expansion, with the result that supposition and possibility suggest almost unending potential. Guglielmo Marconi sought to send a radio signal from Poldhu in Cornwall to Signal Hill, Newfoundland in 1901. Listening through December storms on the hilltop above St. John's, he heard—or believed he heard—the agreed Morse code for the letter 'S', a tiny sound coming through the ether. Subsequent investigations, including those by the scientist, Jack Belrose, of Canada's Research Telecommunications Establishment (DRTE) have created an argument that suggests that this did not actually happen, due to a number of technical factors. Ultimately, of course, Marconi was to achieve his goal, and today's crowded air is witness to the achievement. Yet on that night, whether or not he really heard something, the major significance of the event was that if he did not, then he *imagined* he had heard it.

For the Canadian radio features producer, Chris Brookes, who happens to live at the foot of a cliff immediately beneath where Marconi worked all those years ago, the key factor is that at this point, the medium of radio was appropriated by its own power to inform, through the transmission—or perhaps the *apparent* transmission, of a piece of sonic utility in the form of Morse code. Brookes suggests that had history recorded the Marconi moment differently, as the turning of the key to open a door into the imagination, instead of solely the realising of potential that would transmit facts and information, our view from the start of what radio was, is and can be, might have been different. In Fessenden and Marconi, we have two examples of radio pioneers who broke through the darkness in highly evocative situations. Fessenden's experiment was transformed by the imaginations of those who unwittingly witnessed it, into something extraordinary, with their minds working in partnership with what their senses were telling them.

In Chris Brookes' view, the idea is that Marconi *may* have had a similar experience, based on expectation as he listened with the whole fibre of his being through his headphones, trying to decipher something deep in the fog of white noise: 'Does it give him the information? No. It engages his imagination so powerfully that he *imagines* the information. To me, this illustrates that radio excels not by delivering information (in this case the letter "S") but by evoking the imagination (the *suggestion* of the letter "S"). For radio programme makers this should be a crucial difference' (Brookes in Biewen and Dilworth, 17). If this is an important factor for makers, it should also be a significant consideration for listeners, in how we approach the conscious and active process of attending to sounds. Imaginative radio requires that we respond to it somewhere beyond and beside pure intellect. If the history of the development of radio seems to emerge out of myth, superstition and disputable truths, it is no more than appropriate, given the imaginative roads along which it is capable of transporting us. We would not read a poem in the same way we would read a technical manual, and the accounts of Arthur Burrows, John Reith and others were, as we have seen, full of evocations of the wonder at the root of transmission. Burrows at one point seems to be carried away by his subject, talking of the sun glinting through 'minute globules of dew' which in the eyes are objects, although 'less in size than seed pearls but surpassingly beautiful, are built up systematically of smaller parts, just as St Paul's Cathedral is fashioned from small squares of stone' (Burrows, 2). Perhaps more fittingly for his theme, he suggests 'two prophetic thoughts' as he calls them, which indeed are germane and have been often subsequently cited as images of transmitted sound. The first, spoken by Caliban, in *The Tempest*:

> Be not afear'd, the isle is full of noises,
> Sounds and sweet airs that give delight, and hurt not:
> Sometimes a thousand twangling instruments
> Will hum about mine ears, and sometimes voices…

Burrows ends the quotation at this point, but he might well have continued:

> …That, if I then had waked after long sleep,
> Will make me sleep again: and then, in dreaming,
> The clouds methought would open and show riches
> Ready to drop upon me that, when I waked,
> I cried to dream again. (Act 3, Scene 2)

The other quote Burrows draws from Shakespeare is from *Henry IV*:

> "And those musicians that shall play to you
> Hang in the air, a thousand leagues from hence.
> And straight they shall be here. Sit and attend.'"
> (Act 3, Scene 5, quoted in Burrows: 182)

The *active* engagement of the listener's poetic imagination is transformative as well as collaborative in the process of sonic creation. If there are ghosts in the machine, they are our own creations, invoked by the clues the sounds offer us. There are sounds we imagine inside our minds, our own audio soundtrack evoked by external non-auditory stimuli; likewise, there is the trigger pulled by a single sound or word, as we have seen. Tennyson writes, in his poem, 'The Lover's Tale':

Fair speech was his and delicate of phrase,
Falling in whispers on the sense, addressed
More to the inward ear than to the outward ear,
As rain of the midsummer midnight soft,
Scarce-heard, recalling fragrance and the green
Of the dead spring… (Tennyson, 487, lines 707–712)

Angela Leighton has pointed out that 'the "inward" ear, characteristically, does not so much hear "speech" as listen for something more or less than speech' (Leighton, 59). So it is when we listen to imaginative radio. In January 1924, within two years of the British Broadcasting Company, the producer Nigel Playfair collaborated with the writer Richard Hughes on what was to become the first play specifically written for a listening audience. *Danger* was set in a coal mine, and a collaborative note was struck before it even started, with the station announcer at the BBC's Savoy Hill suggesting that the best effect would be gained by listening to the work in the dark. Thus, with lights off, and the familiar domestic comforts of home removed from the sight, listeners heard:

Mary: [Sharply] Hello! What's happened?
Jack: The lights have gone out!
Mary: Where are you?
Jack: Here.

[*Pause. Steps stumbling*]

Mary: Where? I can't find you. (Hughes, 175)

A radio drama studio is a kind of 'no place', a neutral zone in which words and sound evoke impressions; often—too often perhaps—the impression is that of some actors standing in front of a microphone. In this early experiment, however, by placing the audience in a physical position that created a sympathy with the characters situation, voices from a disembodied source (the radio set) came out of the darkness, and ordinary words, transformed by performance, created a sense of disorientation. Consider also that the audience would mostly have been witness to the broadcast on headphones rather than through loudspeakers, and what emerges is the potential for a total and willing suspension of disbelief. Yet here, in this first written-for-radio play, Hughes succumbs to one of the problems that has beset audio dramatists from the start, an issue that can still undo production intentions to this day, which is to say, the apparent necessity to provide the audience with information and exposition. Jack's line 'the lights have gone out!' is indicative of a dramatist feeling his way as much as does his character within the dramatic situation. In fact, the idea of exhorting the audience to switch the lights out would also, today, be considered unnecessary. The radio drama audience should be given credit for creative listening. They are, after all, a key component in making imaginative radio, and the mind is the screen upon which the maker's pictures are projected. That said, the idea of listening with the lights off has been explored in recent years by a number of groups of radio enthusiasts, gathering to listen in deliberately created darkened environments, tuning in every sense of the word to the power of nuance and suggestion. In 1959, the radio drama producer Donald McWhinnie used the same analogy when demonstrating the nature of the sound radio medium:

> Sit in a darkened room and talk, and listen. Even if you are not vitally interested in words, the words suddenly acquire a compulsion of meaning they did not have before; they develop a richness of texture through being isolated, and you focus your sensibility and imagination on them as you rarely do in daylight. Now play Blind Man's Buff. What voices are coming from which direction? How many feet away? Can this really be an armchair? What is the position of each player related to the next? What is the shape of the well-known, but now hidden, world? (McWhinnie, 23–4)

Thus, an imaginative radio work 'evokes rather than depicts…It is when imagination and emotion begin to colour the spoken word that it in turn begins to have power' (ibid., 43, 51). This notably happens in moments of pause and radio silence, because 'during silence, things happen invisibly, in the minds of the players and in our imagination…sound comes from it, sound returns to it, words have their being surrounded by it, it is the cloth on which the pattern is woven' (ibid.). In the pause after line four in *Danger*, the listener is lost in audio darkness. Then, in her efforts to find Jack, we hear Mary stumble, and we see her again through a sense of her—and our—existence. We feel our way, moment by moment, through things, and 'each moment is calculated in relation to another moment: a total edifice in which words on the page [of a script] and words in the mind, sound and silence on the page, in the mind, in the ear, work together and against each other' (ibid., 102). The voice itself is sufficient to trigger the imagination: the sound of it, the intonation, the ritual of words in a certain, specific order.

Attention All Shipping

Radio ignites unique internal responses in the mind when it acknowledges its own identity as an oral storyteller, because language *was* sound before the two conjoined to make shapes on a page, and there is a deep instinct in all of us to respond to a narrative. It may not be completely linear; the timbre of a voice can carry an implicit story in it, an imagined past, a personality which may be real or illusory, yet which can have the capacity to communicate to a listener who has become a partner in the conspiracy to turn a signal into a relationship. Thus, in communities where the bond between people is strong, and where the continuity of a place is carried through oral history, radio will often thrive best, even in its most utilitarian form. The island of Newfoundland is a case in point; traditionally and historically a centre of fishing since the seventeenth century, its population is scattered into what are known as 'outports' around the coastline, with a comparatively empty and barren rocky interior. There is only one major city, St. John's; for the rest, many communities have their own particular identity and traditions, which have developed through unity of purpose and lifestyle. Such geography develops micro-cultures; Newfoundland's east coast contains a number of families descended from ancestors who came from the west country of England, and in whom the original dialect is still discernable. The south coast villages are strongly Irish in origin, and the west side of the island has a French influence.

In his social history of broadcasting in Newfoundland, Jeff Webb noted that 'research upon the development of Newfoundland popular culture generally, and radio programming in particular, has been pioneered by folklorists, rather than historians. These scholars realised that folk culture was a source of content for mass media and that people integrated content from the media into their folk culture' (Webb, 5). Typical in this was the audience reception and response of a daily programme which over sixty years became an institution. The national broadcaster, the Canadian Broadcasting Corporation (CBC) discovered that *The Fisheries Broadcast*, originally *The Fisherman's Broadcast*, a half-hour programme of news and current affairs relating to the maritime industries around the coast of the island, became an 'appointment-to-listen' for a far wider listenership than its targeted niche specialist audience. People were tuning in for all sorts of reasons not concerned directly with the informative nature of the content. Some said they listened to learn about the island, some to hear the range of voices, others because it gave them a feeling of community and connection, and many because they learned of their country through stories.

As Webb points out, 'radio receivers picked the electromagnetic signals from the air and reconstructed sounds, but listeners had to make sense of them. Discerning meaning from these sounds was a creative act, and the meaning that listeners inferred may have been significantly different from what the originator of the programming intended' (ibid., 9). Today, popular culture of a more homogenised nature dominates Newfoundland. The coastline continues to preserve its pockets of imaginative engagement, and through traditional folk music, the lore continues, linking communities to their past on the island and beyond. Indeed, the marine weather forecast which consumes some eight minutes of *The Fisheries Broadcast* contains implied stories in the very names of its recited coastal regions and sea areas; *Belle Isle, Come-by-Chance, Cabot Strait, Happy Adventure, Great Paradise* and *Fortune Head* seem to speak of hope and opportunity, while *Mistaken Point, Snakes Bight, Wreck Cove* and *Deadman's Bay* suggest stories of another nature to the imagination.

An island such as Newfoundland possesses a historical unity of purpose in its industry which, although dissipated by climate change and overfishing in recent decades, has left a folkloric heritage. This too has diminished as new generations grow away from the past, and in many cases, leave to find lives and work in other countries or in mainland Canada. Yet the blend between the utilitarian and the imaginative persists wherever there is a radio audience. News broadcasts contain more of a sense of ritual, with an emphasis placed (one hopes and supposes) on pure facts and information.

On the other hand, the implication may be transformed by the mind of the audience, given their own personal circumstances, whatever the content; thus some stories do not attract interest, while others engage and animate the consciousness. 'Man is murdered' assumes growing importance in our mind as the location of the event becomes clearer; 'Man is murdered in my town' is clearly significant', while 'Man is murdered in my street' engages us completely. There are, however, exceptions that defy relevance in terms of audience engagement.

Like the Newfoundland Marine Weather Forecast, the UK's daily Shipping Forecast, these days broadcast several times a day on BBC Radio 4, is a litany of names and statistics, directed specifically at a practical and professional audience for whom the information is vital and may be life-affecting. The names themselves—*Lundy, Irish Sea, Shannon, Rockall, Malin, Fair Isle, Faroes*—may not be as immediately evocative as those around Newfoundland, but for some members of the audience they create a complex set of responses; it may be that the longevity of the forecast—broadcast by the BBC on behalf of the Maritime and Coastguard Agency—evokes memories of childhood, because it is probably the longest single daily weather report in history. First created as a working text in February 1861 at the behest of Vice-Admiral Robert Fitzroy (who captained HMS Beagle on the epic voyage of Charles Darwin), it was first published in newspapers in 1867, and transmitted in wireless format from 1911 direct to shipping. The BBC has been broadcasting it since New Years Day, 1924. Thus, it has been a radio presence through several lifetimes. For some listeners, the reasons behind its fascination may be the understated nature of a matter-of-fact emotionless voice, reading names and conditions of wild places to which many of us will never travel. For whatever reason, it is listened to avidly by thousands for whom it has no real significance, nor in many cases for that matter, any functional meaning. It is a fact that has puzzled and inspired many. Composers have set it to music, artists such as Peter Collier have been inspired to travel and endure considerable discomfort to paint the remote locations named in the forecast, while it appears in poetry by Carol Ann Duffy and Seamus Heaney and others.

My own poem on the subject was once chosen by students at a Munich university as a favourite subject for translation; given that Bavaria is completely landlocked, the subject matter would seem nevertheless by this fact to contain an almost abstract imaginative fascination for them. Of its various transmissions through the broadcast day, the reading that seems most potent for listeners is that which takes place just before the station's close down after midnight each evening. Many suggestions have been made as to its imaginative power, among them the cocoon of the warm glow of impending sleep, and the security of the home environment, in juxtaposition with the implied storms raging on the other side of the night, providing the appropriate conditions for what is for many, essentially a meditation or a mantra. The imagination may offer pictures of storm-beaten rocky headlands and pitching fishing boats, but for many it is a poem, the meaning of which lies within itself. As the writer Charlie Connelly has expressed it, 'the shipping forecast turns names and numbers into poetry. There's an innate rhythm to it: it lilts and evokes…Alec Guinness…called it "the best thing on the radio; romantic, authoritative, mesmeric"' (Connelly, 37–8). It has the capacity to take us storm-swept terrains and seascapes through a simple list of names; it is possible to follow the order of the coastal regions as they are spoken, around the shoreline of the British Isles, because the reading proceeds in a clockwise direction: the south coast of England for example is painted in the shorthand of '*Thames, Dover, Wight, Portland, Plymouth*' & c. For anyone seeking to identify the essence of the imaginative partnership between radio and its audience, they need go no further than to listen to an episode of a shipping forecast, be it broadcast on UK radio or any other service.

Opening Doors

Radio as a medium communicates in diverse ways, as a platform for news, information, music, sport, and occasionally—consciously—as an experimental medium, a vehicle for sound art. There are some who would argue that the major networks and commercial companies do not offer enough airtime for impressionistic radio, that from the start, radio was appropriated by the requirement to answer questions rather than ask them. Chris Brookes, sensing, as we discussed earlier, the road not taken in Marconi's 1901 transatlantic experiment, in his search for binary information coming through the ether, suggests the prime purpose behind wireless transmission was identified at that point: 'Unfortunately, on December 2, 1901, the headlines were: MARCONI RECEIVES RADIO SIGNAL! Not MARCONI HAS EVOCATIVE RADIO EXPERIENCE. So we got off on the wrong foot' (Brookes in Biewen and Dilworth, 17). Historically, journalism has predominated along lines of 'need to know' rather than the poetics of enquiry. Charles Dickens in the first lines of *Hard Times* shows it to be a question of temperament:

> Now what I want is, Facts. Teach these boys and girls nothing but Facts. Facts alone are wanted in life. Plant nothing else, and root out everything else. You can only form the minds of reasoning animals upon Facts: nothing else will ever be of service to them. (Dickens 2003, 9)

So speaks Thomas Gradgrind, 'a man of realities. A man of fact and calculations' (ibid., 10). Some sixty pages later, Dickens offers the alternative view:

> It is known, to the force of a single pound weight, what the engine will do; but not all the calculators of the National Debt can tell me the capacity for good or evil, for love or hatred, for patriotism or discontent, for the decomposition of virtue into vice, or the reverse, at any single moment in the soul of one of these quiet servants, with the composed faces and the regulated actions. There is no mystery in it; there is unfathomable mystery in the meanest of them, for ever. – Supposing we were to reserve our arithmetic for material objects, and to govern these awful unknown quantities by other quantities! (ibid., 71)

Imaginative radio, it would seem, has no time for Gradgrind's way of thinking, and through its life it has adopted new technologies wherever it could that could serve as the servant of the imagination. For example, the concepts that were to develop into the famous BBC's *Radiophonic Workshop* had already, as they began to emerge, created other sounds that complemented text in early works such as Samuel Beckett's *All That Fall* (1957) with a logic that shifted the sound world from moment to moment, embedded even as it was in text. In Donald McWhinnie's production, it becomes a fundamental part of a drama in which, by 'reducing the sensorial spectrum to pure sound, Beckett is able to move beyond language – indeed, revel in the "bizarre"-ness of it – to explore the way other kinds of sound impact the total experience of the play' (Niebur, 20). On occasions such as this, we are being seduced and manipulated, and we willingly accept it, stepping into the strange world of characters such as Maddy Rooney, and going where these shadows take us, 'halfway between waking consciousness and fantasy, even hallucination. Just as the subject experiencing such states finds it difficult to decide whether his experience is real or fantasy, so the radio dramatist can keep the listener in a similar state of uncertainty' (Esslin, 131). That said, our listening is actively creative; we build from the inside out, shaping impressions from accidental encounters with sound that brushes us tangentially on its way past us. It is a fact that much of the time, the journeys upon which we embark in the sonic imagination are not only rendered remarkable by the way things are made, but by the way we interpret them and respond to them. Everything from the final point of production onwards is a matter of sharing. Whatever and whoever is the agenda, the result of active listening to sound can be a profoundly immersive one, be its source consciously created as art or narrative, or the result of an imaginative interpretation within ourselves of the information we absorb.

As radio has become more polarised between news and popular music stations, and the artistic freedom enabled by the development of pod-casting, the emergence of sound art has appropriated the technology to create works that use the medium as it were, as a character in itself. In the next chapter, we shall explore through a case study the famous example of Orson Welles's 1938 CBS broadcast of *The War of the Worlds*, which interrogated both the technique and effect of what was still at that time, and emergent form. The Canadian, Anna Friz has since 1998, predominantly created self-reflexive sound works for broadcast, installation or performance where radio is the source, subject and medium of the piece. An example of this is *The Clandestine Transmissions of Pirate Jenny*, a work that has grown in a number of ways across a range of media platforms. The central character, Jenny, is a voice trapped inside the physical entity of a radio set, at the mercy of the mass media that passes through the equipment and which she is forced to process. Plaintively speaking into our ear, as if fearing she will be overheard, she sends us desperate messages for help, stolen moments between her slavery to what contemporary radio has become. Starting as a fantasy of the memory of innocently wondering, as so many children once did, about the little people who lived inside the radio, and how they got there, the drama—for drama it is—grows darker, becoming a study in alienation and loneliness, a comment on modern broadcasting and a powerful picture of human isolation. What makes it remarkable, years after it was first created, is not only its continuing—indeed, increasing—relevance, but the fact that it is built out of the very blocks of technology itself, simulated old-style radio, valves, tuning, static and all, as if we have come across Jenny and her plight accidentally, while scanning through distant broadcasts. Yet it is experienced as often as not as a download, or in a performance space. In so doing, Friz interrogates and subverts the actual technical vocabulary of the medium she is using as a vehicle, just as did Orson Welles in 1938 on CBS. As is so often the case with new advances in technology, forces seized upon by experimental 'fringe' creators become appropriated by mass broadcasters and developed with funds unavailable to the pioneers. It is a hope and wish that more funding can be found for innovation on the edges of media, because this is where the truly 'new' is so often located.

New technologies of dissemination and listening provide alternatives to the traditions that mainstream radio has developed for itself. Community stations and niche arts broadcasters offer more open policies for poetic and artistic making, and the development of podcasting provides ways in which voices frustrated by lack of access to transmitted radio can speak with new and exciting individual voices, and the Internet is for many, the ultimate form of artistic democratisation. As Sabine Breitsameter perceptively underlined some years ago, 'these new possibilities may open up surprisingly new audio "visions" and provoke new strategies of perception. Audio on the Internet has made the boundaries between art, communication and play flexible' (Breitsameter, quoted in Gilfillan, 6[1]). Breitsameter's prophecy of 2003 is today manifest in the freedom of a multitude of young sound poets, working with minimal facilities to create some of the most interesting sonic works, for which the word 'radio' is suddenly not quite adequate.

There is an analogy to be found between the way we listen to imaginative sound and its origins in production. When radio was created, listening was enabled through headphones, because speaker technology was not equal to the task of filling a room with sound. Even prior to this, a device known as the Electrophone available between 1895 and 1925 and based on the earlier French Théâtrephone utilised the telephone line to pipe live public events such as church services and concerts, into the home, to be heard by holding the receiver to the ear. With the coming of audio speakers came the idea of sharing the experience, so a play or a story could be communally appreciated, commented on and indeed, if the attention wandered, talked over. This is in itself an aspect of the very idea of broadcasting from a transmitter, in that 'broadcasting', an archaic term derived from the wide sprinkling of seed was a haphazard dissemination of information, some of which might bear fruit, while some might fall on barren ground.

Today's podcast culture is linked to individual active choice and personal listening, and perhaps to a sense of isolating ourselves in the consumption of our personal media experiences. We do not tune in: we download, save an item and then decide where and when to listen. It is an intimate experience, made even more so by the means by which a large part of the listenership will absorb the audio event, that is to say via headphones or earbuds. In this case, the sound is not something heard across a room, a second-hand experience, absorbing as it comes to us the acoustics of the space in which it reverberates; Now, it is literally talking into our ears, wherever and whenever we choose to access it. As Spinelli and Dann have pointed out, 'We could say…that ear-buds allow for a hyper-intimacy in which the voice you hear is in no way external, but present inside you. Ear-buds push intimacy inside a body – they are, in a very real sense, about re-*embodying* the voice. This observation cuts against a discourse familiar to media writing for nearly a century which described radio as a "disembodied" voice' (Spinelli and Dann, 84).

And yet, irrespective of technology, the true innovator remains the human imagination, and we each of us possess one; the most prosaic of factual utility may provide a flight of fancy, be it a phrase, a pause that conceals a sob, or even a set of statistics heard later at night, taking us to a shoreline of wheeling seabirds, crashing waves inhaling shingle and the unknown potential of weather, held in the barometrical words, 'now falling'. Whether it be old-style radio or sound communicated through innovation, we are storytelling creatures, listening and reinventing our-selves through the accumulation of ideas and dreams, and sound was there at the start of the very first story. It is complicit; unlike visual media, it does not supply pre-packaged information in an instant, but builds through moments of time and sparks of imaginative ignition. It is linear, temporal; we might almost suggest that sound has its own mortality, subject to the same laws as those who live through it. Meaning gradually emerges, falls like rain into the consciousness, ebbing and flowing like tides, withholding and revealing as it does so. Today, we can choose it by downloading, or it can choose us through serendipity, surprise or shock us into insight or inspiration when we least expect it. Whatever the selected means of absorption, be it transmitted radio or downloaded podcast, sound has the capacity to change us forever, because we can never know, at first listening, what the next sound or word will be. Audio in all its forms can deliver the shock of the new in ways that a listener will readily accept and embrace.

We have heard what memory offers us as to the possible sounds of history, and we build our experience of life—and our responses to that experience—upon what we know. Likewise we have located within ourselves the limitless silence of possibility, within which these things are stored. It is from here a short step to a point where we can hear the sounds of tomorrow or of other shadowy parallel worlds such as those occupied by the trapped characters given voice by Anna Friz. Whether or not such places exist, or will exist as we imagine them, we might argue that they are already happening from the moment we invent them as potentials. If we can travel instantly to 'Sea area Malin Head' through the suggestion of a phrase in the shipping forecast, where else might we go? As we shall see in the next chapter, this is the key to doors opening into strange places beyond us and within us.

Note

1. Brietsameter, Sabine. 'What Is AudioHyperspace?' *Audiohyperspace—Akustische Kunst in Netzwerken und Datenräumen.* Süwestrundfunk. www.swr2.de/audiohyperspace/engl_version/infoindex.html.

6

The Sounds of Shadows and Light: Science Fiction and the Supernatural

Sleepless Imaginings

Just as every story begins in the imagination with the phrases 'what if…?' and 'supposing that…?' so do the prisms of sound emerge from silence, brushing past us, transient but beyond our capacity to 'un-hear'. Look at a tree. Blink and look again. The tree remains much as it was a moment ago. The sound of what it was just now, however—the breeze in the leaves, the creak of branches, the rustle or song of a hidden bird—these are now forever changed. Sound is a ghostly, fleeting presence, and some of the most potent sounds that affect us are the symptoms of things that do not exist in reality at all, or which are at least unexplained. Freud wrote of 'das unheimliche'—the uncanny—and it is perhaps true that today we take for granted sounds and events that we might once have considered strange or incomprehensibly alien. Yet there remain places of deep superstition, around us and within us. Points in a landscape have their own character-istics, be it a bright acoustic, a timbre, a rhythmic sound, passing animal noises or birds, and occasionally something unidentifiable. Making any sound recording and then playing it back at a later date, away from the original location, can have a powerful effect on the imaginative memory,

© The Author(s) 2019
S. Street, *The Sound inside the Silence*, Palgrave Studies in Sound,
https://doi.org/10.1007/978-981-13-8449-3_6

often much more so than a photograph or a film. There is sometimes too, an intangible feeling, difficult to explain or quantify, a feeling of being in a unique environment with its own peculiar atmosphere. We simplify it when we speak of achieving in our recordings 'a sense of place'.

For his CD *Into the Dark*, made in 1994, Chris Watson explored such places, identifying locations through local natural or social history research, reading maps or from stories, folklore, conversation and anecdote. As an extra guide, he was aided by the researches of the writer Tom Lethbridge who identified the sources of several 'spirits' within the topography of the area. Out of this came the cries of rooks in Embleton churchyard, said to be the reincarnated voices of drowned sailors. A location may seep into our imagination, and some atmospheres insinuate whispers and murmurs to which we may assign our own meanings, or which impose their own on us. Watson walked through a merciless wind in a snow clad Glen Cannich with his microphone, feeling himself to be a small part of a giant sound instrument, and at Gahlitzerstrom, Ummanz, on the Rügen Islands in Germany, where the beat of Cranes wings reminded him that in Greek mythology, where Hermes was said to have conceived the Greek alphabet by watching the beating wings of Cranes as they passed by his line of sight. In such places, we come to the edge of myth and legend, and they seem to envelop us with their natural music. Aware of the stories, we hear the sounds and willingly believe. Even without the narratives, these recordings haunt us; indeed heard as pure audio without, as it were, a programme note to tell us how we should feel or think, the sounds bring their darkness and strangeness into the space of the imagination, and by entering it, turn it dark too.

A natural sound, however atmospheric, while it may chill or startle us, might at least be something we can identify, but the sound that does not have an apparent reason, or the sound we invent in our own mental recorder, has the power to unleash a deeply imbedded sense of unease within us. It is this response that film-makers seek to release when they simulate the cry of a dinosaur or a monstrous form from a haunted house or a distant galaxy. Digital sound and mixing can be kaleidoscopic, but as the sound editor and designer Walter Murch once said: 'ideally the perfect sound film has zero tracks. You try to get the audience to a point, somehow, where they can *imagine* the sound. They hear the sound in their minds, and it really isn't on the track at all. That's the ideal sound the one that exists totally in the mind' (Murch in Weis and Belton, 359). Beyond even this is a sense of how the imagination implies things suggested by what the ears receive from the sounds that *are* present. We draw conclusions from available evidence, or from the lack of it.

Matching this imaginative idea is what fuels creative technology in so many science fiction films and computer games. Sound media such as radio have less work to do, because there are no physical pictures to 'give the game away'. The English poet Thomas Hardy suggested that silence is the most eerie and multifaceted sound of all, possessing a myriad of meanings and possible interpretations. We hear a creaking floorboard in the shadows, and our mind creates its own 'dark night of the soul'. The strangeness of radio, discussed in the previous chapter, comes back into play here. If sound signals travel through the atmosphere, where do they go, and who do they reach? The wireless pioneer, Oliver Lodge, believed that this technology had the capacity to communicate with the dead, and the same idea fuelled the researches and experiments of the Latvian Konstantin Raudive, who claimed through many years of concentrated effort to have recorded numerous examples of voices from deceased individuals. In some senses, this is akin to the uncanny nature of theatrical ventriloquism as practised in Victorian and Edwardian theatres and variety halls, through which the artist by throwing his or her voice around the building could induce states of near terror in an audience by the very strangeness of sounds emerging from apparently empty space. It may be argued that the truest recording of a loved one may forever remain unplayable because it lies in the mind: within the memory and imagination. It is hardly surprising that technology as it evolved and contributed to the field of aesthetics would be used as a complement to studies of artistic expression, as in the work of the futurist, Luigi Russolo, the Italian painter, composer and technologist, who underlined the occult as the foundation for his own art. Understanding the root of an existential fear of being alone in time and space is at centre of the idea of much supernatural and science fiction sound design. Here, where often less is more, and an almost-stillness is far in excess of a roar when it comes to engendering fear, we confront ourselves.

Fact is a thin layer atop a deep and mysterious iceberg, and for many, the very phenomenon of invisible signals travelling through the air has always been akin, as have seen, to the miraculous. Indeed, inspired by its very strangeness, some have gone further, seeing no limit to the potential for communication. This chapter explores this from two perspectives: firstly, we shall examine the role of some investigators into the realms of parapsychology and connections to be found in occult, theosophical, futurist and spiritualist studies, and the significance of voice and music in this area. This is sound as suggested evidence, documented to be believed or rejected. Secondly, we shall discuss the field of science fiction as it seeks to capture the attention of mass audiences, and to explore as it does so through metaphor some of the social issues of our time.

Ether and Ectoplasm

In 1971, the Latvian scientist, Konstantin Raudive, wrote a book, published in the English-speaking world as *Breakthrough*. Based on years of research, and tape recordings purporting to have been made under laboratory test conditions, he claimed to have recorded voices transmitted from another dimension. Working with the German parapsychologist, Hans Bender, Raudive, who was also a practising Roman Catholic, gave his project the name of electronic voice phenomenon (EVP). Prior to this, as early as 1936 an artist named Attila von Szalay claimed that, while working in his darkroom, he heard the voice of his dead brother, prompting a twenty-year quest to secure recordings of what were thought to be spirit voices. Subsequently, in the 1950s a Swedish painter Friedrich Jürgenson began a series of experiments based on what he claimed was an accidental discovery. Jürgenson, who was also a documentary film-maker, reported that while playing back recordings of some birdsong at his window, he heard the voices of his dead father and wife calling his name. Raudive read Jürgenson's book of his experiments, *Radio Contact with the Dead*, published in 1967, and utilised some of his techniques for recording.

The two men began a collaboration, but initially their efforts were largely unsuccessful. According to Raudive, however, one night, as he listened to one recording, he clearly heard a number of voices. When he played the tape back repeatedly, he began to detect individual words and phrases spoken in various languages, Latvian, German and French. The last voice he heard, that of a woman who called herself 'Margarete', made a strong impression on him, believing as he did that this was one Margarete Petrautzki, someone known to him, who had recently died, and for whom he had some affection. Thus began his work in earnest, and with the help of various electronics experts he recorded over 100,000 audio tapes over the coming years, most of which were made under what he described as strict laboratory conditions. Over 400 people were involved in his research, and all apparently heard the voices.

In his writings, he documented meticulously the case studies that arose from the work, and also some of the insights gained from the speakers themselves. One of the most startling was the claim that spirits in the afterlife had technologies—'equipment'—enabling contact with the material world:

> The astonishing conception that "other-worldly" transmitting stations exist, emerges quite clearly from many of the voices' statements. Information received indicates that there are various groups of voice-entities who operate their own stations. The experimenter has recorded this phenomenon on hundreds of occasions and has submitted the tapes to experts for listening-in-tests. (Raudive 1971: 174)

Raudive sought to explain how, in a non-material world, such things were possible: '[They] do not only use their own transmitting and receiving stations, they also have to apply their own special type of electronic technique' (ibid.). What makes the study of Raudive's experiments in sound a melancholy one is the overwhelming sense that, if these voices were to be proved genuine, they would clearly be vestiges of personality lost in some strange netherworld of white noise, as Jeffrey Sconce has written, an infinity of 'continuing strife and the terrifying phenomenon of interstitial uncertainty, a disassociation of mind, body and spirit that seems impossible to reintegrate' (Sconce 2000: 88).

Our media offers diverse possibilities; we may believe, as Raudive did, that this is the evidence of true communication between material and spiritual worlds, not imagination at all, and so seek to generate a dialogue.

On the other hand, we may consider all this to be material from which to make stories, playing on our deep superstition and fear of what might lie in the darkness at the top of the stairs. In so doing, we may indeed find these voices to be occupying the haunted house that is our own psyche. The radio that is our mind plays out our hopes and our fantasies, and as we know, one of radio's most significant and powerful qualities is its ability to create tension and expectation through potential in the listener's mind. Silence, a pause, a sudden stillness, alerts the sense to what might happen next; the brain jumps forward to generate a possibility. It is why ghost stories work so well in sound alone, just one step away from the printed page. Writing of her location-based drama, *Ropewalk House*, broadcast on BBC Radio 4 in January 2019, the playwright Anita Sullivan said:

> You'd think "waiting" on radio would be dull, but suspense is about listening, holding you in that moment of stillness before everything changes. Radio does that "must-be-silent" journey through the cellar, attic or creaking house so well. And it isn't the character listening, trying not to breathe, it's *you,* waiting for flight or fight, or for everything to be alright.[1]

The moment *before* the event carries its own unique and particular potency.

*   *   *

What makes Raudive and similar researchers of interest in the context of this book is the fact that here we are confronted with claims that what had been previously considered to lie in the field of storytelling, imagination, dreams and superstition, could now be demonstrably proven with hard evidence. It is not the place of this study to argue for or against these claims. I have placed the supernatural together with fictionalised explorations of science only to demonstrate that here we enter unknown regions that are of perennial fascination across all races and religions. Both fields of research and narrative feed off the belief among many that, other worlds than ours may exist, could we but identify the means to recognise them, and then to communicate. These may be malign phantoms, benign spirits, angels or aliens. In sound terms, all these areas have been fruitful throughout the history of conjecture and storytelling, from the time of bards and troubadours to the present day. For the sound designer and the audio producer, the unknown opens doors to limitless fields of expression, if only because anything is possible.

Sounds of Hope and Possibility

Like Raudive, the British physicist Oliver Lodge (1851–1940) was a scientist who saw infinite possibilities in the new technology of radio telegraphy. He was a pioneer experimenter with wireless and perfected the coherer, a radio-wave detector which was to be at the heart of early radio telegraphy receivers. His academic career was a glittering one. From an assistant professorship at University College, London in 1879, he was appointed as Chair of Physics at University College, Liverpool in 1881, when he was just thirty. While in Liverpool he conducted experiments in the propagation and reception of electromagnetic waves. When in 1890, the French physicist Edouard Branly showed that loose iron filings in a glass tube coalesce, or 'cohere', under the influence of radiated electric waves, Lodge was inspired to improve his own invention, enabling it to detect Morse code signals transmitted by radio waves and enable them to be transcribed onto paper. Lodge's device, first demonstrated before the Royal Institute in 1894, quickly became the standard detector in early wireless telegraph receivers, although outmoded after the turn of the century. Lodge also obtained patents in 1897 for the use of inductors and capacitors to adjust the frequency of wireless transmitters and receivers. In 1900, Lodge was chosen as the first principal of the new Birmingham University, and he was knighted in 1902. Then, from 1900 onwards, he became prominent in psychical research, believing strongly in the possibility of communicating with the dead. This subsequently became an intensely personal quest when his son Raymond was killed in the First World War, and his attempts to communicate with him are movingly described in his book, *Raymond*.

Key to much of Lodge's studies, particularly at the crossing point between science and philosophy, was his consideration of the mysterious material known through history as *aether* or ether. According to ancient thought, ether was a substance that filled the universe above the earth's atmosphere. The concept of ether had been used in several theories to explain certain natural phenomena, such as the travelling of light and gravity. In the late nineteenth century, physicists postulated that ether permeated throughout space, between stars and planets. As we have already heard, John Reith and his fellow radio pioneers voiced a fascination for the subject; in this sense, and in the immediate post-First World War years, Lodge was a man of his time, and there were many bereaved families who were only too willing to accept his findings, as he developed them.

Lodge had referred to Newton's enquiries into ether in an 1882 lecture delivered at the London Institution that same year. Even then it seems likely that Lodge was making connections beyond physical science. For him ether 'was already notable for its exceptionally abstract quality and very probably drew some of its inspiration from sources which were anything but "physical" in origin. Originally, these may have stemmed from an orthodox religious position…' (Rowlands, 14). Later, however his explorations led him 'into the much more murky waters of psychic research, the "paranormal" and spiritualism' (ibid.). During his time in Liverpool, Lodge became the founding president of the Liverpool Physical (not to be confused with Psychical) Society, and accounts of its proceedings held in the archives of Liverpool University show that in his research, nothing was off limits: for example, in one entry, we find him suggesting the possibility of detecting radio waves from extraterrestrial sources: 'I hope to try for long waves radiation from the sun, filtering out the ordinary well-known waves by a blackboard, or other sufficiently opaque substance'.[2]

After his retirement, he continued to research into the ether, seeking to make connections with other scientists to gain support for his ideas, which he persisted in believing to be key to the universe. Unlock the secrets of the ether, Lodge thought, and everything else will be explained. 'He suggested, indeed, that there was nothing in the physical universe but the ether, in various states of rotational motion and carrying all manner of periodic waves. A small portion of the vast store of energy in the ether was identifiable as the ordinary particles making up the material universe. From interaction of matter with the rest of the enormous amount of energy arose the phenomena not yet understood: life and the mind' (Jolly, 230). In 1933, Lodge published *My Philosophy: Representing My Views on the Many Functions of the Ether in Space*, the object of which was to show that 'there is an etheric world which spreads over both regions of Religion and

Science, and may be the means of reconciling them...' (ibid., 236). In a key passage, he writes:

> The revolution in physical science which has been going on through this century has had the effect of directing our main attention away from material bodies and concentrating it on the multifarious happenings in space. In my belief this process will go further, for that is where our real existence lies, and there is our spiritual home. It is in the interaction of ether and matter that the problems of psychology must find their solution. Our existence is essentially independent of this material organism that we have constructed and use for a time. Matter only serves as an index or pointer, demonstrating the unseen activities are the etheric agencies guided by Life and Mind and Spirit. (Lodge, 132)

Lodge saw his researches into the new invisible medium of wireless technology as entirely complementary to his ideas of the survival of death. In order to understand his motives and thinking, we must place ourselves in his time and engage imaginatively with the possibilities suggested by his technical discoveries, offering as they did for so many, new possibilities, and in the face of terrible loss, hope for the transportation of the sound of ideas and identities between worlds. Oliver Lodge and Konstantin Raudive, had they been contemporaries, would surely have seen one another as allies.

*   *   *

Sound and its invisible communicability have always struck chords of connection within the human imagination, be it in science or art. Luigi Russolo was considered by many to be one of the first noise music composers and practitioners. Russolo's view was that traditional melodic music was confining, and he saw noise music as its future replacement. In 1913, he published his manifesto, *The Art of Noises*, a manifesto that called for a new way of relating noise and music. Russolo sought to create a new music that moved away from the man-made instrumentation and forms of history.

Let it follow the line of the rainbow and vie with the clouds in breaking sunbeams: *let music be naught else than Nature mirrored by and reflected from the human breast;* for it is sounding air and floats above and beyond the air; within Man himself as universally and absolutely as in Creation entire; for it can gather together and disperse without losing its intensity. (Russolo, 98)

It is impossible to consider one aspect of Russolo's thinking without allowing it to embrace a cross-referencing of all art, which in his mind was increasingly informed by the common denominator of the occult. Among these concepts were synaesthetic ideas derived in part from the symbolists and the scapigliati,[3] in turn shaped by occultist theories; 'The perception of all the arts as secretly linked by the theory of vibrations allowed Russolo to move freely between them' (Chessa, 112). Key to his practice was his invention of musical instruments called the 'Intonarumori' or 'noise-intoners', which he used to replicate the sounds of the industrial age. 'For Russolo, the Intonarumori were part of an alchemical experiment into the creation of life, which futurists believed was the only process capable of producing art that could truly be called "spiritual." In Russolo's experiment, raw matter (in the form of pure noise) is transformed by means of mechanical instruments (the Intonarumori) functioning as an alchemical crucible' (ibid., 138). Russolo, in parallel with the futurist movement of his time, fully accepted that scientific and spiritual goals were mutually compatible, something, according to Chessa, that was missing in much thinking after the Second World War regarding the work: 'Once the general interest in theosophy had waned, modernist criticism of Futurism entirely missed the futurists' equation "Occult=Science"' (ibid.).

By the mid-1930s, however, with the development of new technologies, notably the media, the growth in materialism and communal paranoia regarding the state of the material world had to a large extent shifted popular thinking away from fascination in such subjects as theosophy and spiritualism. Cinema and radio continued to point outwards from the earth, but it was science fiction, and communication not with the dead, but with other galaxies, that seized the imagination, both as escape and metaphor. The very fragility of communication, be it with supernatural or intergalactic beings, and questions as to the benignity or otherwise of these forces stand as the bridges between investigation—scholarly, artistic or casual—and fiction; when the few facts we have fall by the wayside, the storyteller takes over, and in these fields in particular, sound is the perfect medium, because, like the written word, it partners rather than competes with the imagination. 'The interpretation of sound as an unstable or provisional event, ambiguously situated somewhere between psychological delusion, verifiable scientific phenomenon, and a visitation of spectral forces, is a frequent trope of supernatural fiction' (Toop, 130). The investigations of Raudive and Lodge and the fictionalised 'entertainments' that relate to them are linked because while one seeks to *document* the unknowable, the other exploits it imaginatively. Likewise we may suggest that whatever the worlds with which we may seek to communicate, be they within us, an etheric world of spirit or somewhere in the outer cosmos, in our searching, we are touching on the ancient idea of the Music of the Spheres, a concept—with its roots in Pythagoras—that places sound at the centre of the universe. The idea that there comes from the stars, planets, suns—even from the space between them—a resonating harmony too powerful and of a frequency too subtle to be perceived by human hearing, has energised the thinking of composers from Purcell to the present day. It is indeed an extra-musical idea that enters philosophy as well as music.

While science fiction in some of its unearthly manifestations may touch on the supernatural and spiritual, insofar as it may portray utopias as exemplars of ideal living, it also plays its part in politics of a more physical nature, in particular with its potential for use as a metaphor, as in the case of the McCarthy hearings in the USA during the 1940s and 1950s, when aliens from outer space became the analogy in sci-fi movies for worldly paranoia. Radio, however, has always had a more direct communication with the imagination, and creative producers have utilised this quality since its early years. Be it the propensity to look within itself for meanings, as in the solitary soliloquising of Anna Friz's *Pirate Jenny*, discussed in the previous chapter, or in the popular adventures of such radio space heroes as Buck Rogers, Dan Dare, or Jet Morgan and his crew in the BBC series of the 1950s, *Journey into Space*, written by Charles Chilton, it is the technology with which we surround ourselves, or which we can imagine, that is the trigger. It is full of possibilities as limitless as the imagination of the writer or producer, and 'wherever streams of consciousness and electrons converge in the cultural imagination, there lies a potential conduit to an electronic elsewhere that, even as it evokes the spectre of the void, also holds the promise of a higher form of consciousness, one that promises to evade the often annihilating powers of our technologies and transcend the now materially demystified machine that is the human body' (Sconce, 92).

Alien Invasions

In 1938, Orson Welles used these criteria in conjunction with other factors in the production of his CBS radio broadcast adaptation of H. G. Wells' novel, *The War of the Worlds*. The story behind this infamous event has been told and retold many times and has even been the subject of a film, but it warrants some re-examination and discussion in the context of this exploration of the auditory imagination. Here, the world of the unknown offered not hope of survival, but the fear of invasion and the mass destruction of civilisations. Playing on the audience's familiarity with the role of the station announcer, and the broadcast reportage of journalists, the CBS drama gradually built a sound picture to support evidence of attack from another planet and the subsequent breakdown of society. Although the on-air process from initial threat to final annihilation took little more than half an hour, nevertheless it created a suspension of disbelief sufficient to create a broadcasting phenomenon that continues to be analysed to this day. Indeed, in an era where the term 'fake news' has become familiar, and the truth behind current affairs has seemed to grow more illusive, the Orson Welles broadcast remains highly relevant. The programme was divided into two slightly unequal halves, separated by a commercial break. Clearly, a sponsor's message coming where it did would end any credibility for purported truth, if such was in fact intended. Thus Welles and his writer, Howard Koch, chose to end the first part of the drama with the bleakest possible radio event, the destruction of the network itself, followed by the voice of a single DXer, calling out into silence, sonic darkness, desperately seeking another human voice by way of response:

2X2L calling CQ….
2X2L calling CQ….
2x2L callin CQ….New York.
Isn't there anyone on the air?
Isn't there anyone….
2X2L ——————— (Cantril, 31)

During the autumn of 1937, the Rockefeller Foundation allocated a grant to Princeton University (from which in the radio adaptation, the fictional Professor Richard Pierson, played by Welles in the broadcast, was supposed to have come[4]) to explore the affect of radio on American listeners. An Office of Radio Research was set up, directed by Frank Stanton and Hadley Cantril, and a wide range of case studies was explored. 'In the fall of 1938 the broadcast, *The War of the Worlds,* by the Mercury Theatre, provided an unexpected "experimental" situation. A special grant by the General Education Board made it possible to study the event which fitted so well into the whole frame of the Princeton Project' (Cantril, v). Further studies were sponsored by the Federal Radio Education Committee, resulting in a book, *The Invasion from Mars: A Study in the Psychology of Panic*, which was published in 1940, going through a number of subsequent editions.

Taking into account the zeitgeist of the time, in an age before the Internet and instant social communications, it becomes easier to understand how a news event pretending to come from a supposedly credible source—CBS Radio—might have persuaded a section of the listening audience of its veracity. There was one other major broadcaster, NBC, to which the majority of the audience would have been listening at the time—8.00 p.m. on a Sunday evening—and this was broadcasting a variety programme, *The Edgar Bergen/Charlie McCarthy Show.* NBC carried a sponsor's message a short way into their show, and at this point, listeners who chose to retune would have heard CBS broadcasting a weather forecast followed by a programme of popular dance music, apparently from 'the Meridian Room in the Hotel Park Plaza in downtown New York', by 'Ramon Raquello and his orchestra' (impersonated by studio musicians led by Bernard Herrmann). The first piece was 'a touch of the Spanish, "La Cumparsita"'. The original had been recorded in 1930 by Carlos Gardel, a French Argentine singer, songwriter and actor. He was acknowledged internationally as a major figure in the history of the Tango, and although he had died in 1935, three years before Bernard Herrmann appropriated it, his song was still popular, and the familiarity might well have persuaded the casual listener to remain with the network. Had they done so, after a minute or so, they would have heard the infamous words from the studio 'announcer', 'Ladies and gentlemen, we interrupt our programme of dance music to bring you a special bulletin from the Intercontinental Radio News' (Cantril, 6). Thus, fiction seamlessly merged with fact, and an intrigued audience might have been interested enough to remain, rather than tuning back to the variety show, and so be pulled into the deception, having missed the opening explanatory announcement that this was in fact the latest in a series of dramatic presentations by Orson Welles's Mercury Theatre company.

Much has been made of the question as to whether the effects of the 'Martian Broadcast' on its audience were as powerful as contemporary accounts have led us to believe. One view is that 'the vast majority of listeners understood the broadcast correctly…The "panic", as many newspapers defined it, only began when some listeners passed the fake news on to unsuspecting others, spreading their fear and confusion' (Schwarz, 223). That said, in a country the size of the USA, it would only require a small ratio of 'believers' to register a significant response within the listening audience. The Princeton Project commissioned a questionnaire which it circulated through two hundred and fifty social psychologists and sociologists, asking them to consider the characteristics and concerns of people within the context of the time who might have been most likely to have been frightened by the broadcast. Measured on a scale from nine to zero, where nine was the most powerful justification and zero being the least, the factors were as follows:

8/9: The recent war scare in Europe.
7: General intellectual immaturity combined with a belief in the prestige of radio announcers.
6: General emotional immaturity.
5: Science as a mystery.

Less influential in audience response was:

> 2/3: Insecurity from prolonged social depression combined with insecurity from natural catastrophes.
> 1/2: Belief that the world will end sometime, combined with reading of "Buck Rogers," etc. [in other words, thinking influenced by popular science fiction.] (Cantril, 211)

Bottom of the list, although still just measurable, came evidence of responses based on religious beliefs. Whatever the degree of susceptibility to deception it is clear that the power of suggestion, carefully engineered in this case and taking into account elements known to be true, could plant a seed with the potential for rapid growth. (For example, such was the international situation at the time, that announcer interruptions with newsflashes had become a convention readily understood by the listening audience.[5]) After the broadcast, Welles distanced himself from accusations that the deception had been deliberate, although he later admitted that part of the intention had been to test the developing medium of radio, and the audience's response to it. (The relationship between media and power was to gain further expression from him in the film of *Citizen Kane* in 1941.) He, and his writer, Howard Koch, had been made aware—as had much of the listening public—of the power of spontaneous radio in Herbert Morrison's commentary on the crash of the airship Hindenburg at Lakehurst, New Jersey in May 1937. This would have been still fresh in the minds of audience and makers by the time the broadcast was produced. The idea of unplanned and unscripted events overtaking the polished professionalism of the broadcaster, as demonstrated graphically in the Hindenburg transmission, would have been a major influence on how *The War of the Worlds* was scripted.

   Given the relatively primitive facilities available—in a live studio—to make the Martian broadcast, transmitted as it happened, it is understandable that to modern ears there may be some somewhat crude production elements in the execution of the work. Nevertheless the intention behind it, and the effect, remains impressively persuading, particularly if listened to as background. Most members of a radio audience hear broadcasts more or less filtered through a gauze of everyday life, and while content may not always be grasped, certainly pitch, timbre, shades of alarm in voices and so on will usually register in some form within the conscious or sub-conscious mind. Certainly, even today, any sense that the broadcaster may not be in control of the situation will always attract a sharpened focus of attention. On that Halloween night in 1938, public awareness of the crisis in World affairs, emergence from financial and social Depression and an overall sense of public anxiety, together with the skilful manipulation of sound forces, conspired to plant imaginative tremors that continue to reverberate and fascinate to this day. Listening to the play now, the audio sepia of an archive recording invests it with a curious sense of authenticity, as if strangely, it may during the intervening years, have actually come true after all. *The War of the Worlds* remains a potent illustration of the power and subtlety of sound as a direct communicant to the human mind, with all its insecurities, phobias and imaginative capacity for invention, ignited from small flickering flames of suggestion.

*   *   *

In the previous chapter, we touched on the contribution of electronic sound to radio production, as used by Donald McWhinnie in Samuel Beckett's first radio play, *All That Fall*. Through the late 1950s and the early part of the decade that followed, creative composers working within the BBC used new and developing technology to pioneer sounds that complemented some of the more experimental work being explored by writers and producers. The *Radiophonic Workshop* developed music and sound effects that could not be created on man-made instruments, taking radio and television fantasy sound into new realms. Yet it should be remembered that this was part of a continuous process that had begun long before, as producers of radio dramas and features sought to find fresh ways of expressing narratives involving the use of sounds other than words, familiar studio sound effects and conventional incidental music. The use of new technology, that is to say sound/music produced by non-conventional instruments and therefore distanced from their human 'players'—provided imaginative links with the supernatural, with magic and with science fiction. McWhinnie spoke of "'a science of making sound patterns" capable of creating "a vast and subtle symphony from the sound of a pin dropping…A sort of modern magic'" (Niebur, 29).

Graphic Radio

In 1996, the British producer Dirk Maggs created *Independence Day UK* , a parallel audio production based on the American science fiction film *Independence Day*, an unfolding story of alien invasion. When it was broadcast on BBC Radio 1, with known radio journalists commentating on action that seemed to be developing minute-by-minute in the sense of a live current affairs broadcast, it conveyed something of the blend of fact and fiction that had so disturbed listeners to the 1938 broadcast of *The War of the Worlds*. Maggs made use of cutting-edge digital technology to create a sound that was the sonic imaginative equivalent of watching a three-dimensional film, and it was this characteristic that gave many of his productions a quality that offered radio drama new possibilities of expression. Indeed, through the employment of this technology, combined with cinematic-style music, multi-layered sound effects and the use of voices that were treated and enhanced to give them a heightened presence, he developed a form of radio drama that justly earned the epithet of 'audio movies'. From 2003 to 2005, Maggs also produced new episodes of *The Hitchhiker's Guide to the Galaxy*, based on books written by the late Douglas Adams, in which Adams had pursued ideas from his groundbreaking comedy science fiction series, first broadcast in 1978. The programmes had gained cult listening status, partly because of the idea of giving alien beings human qualities and failings, but largely because of the innovative sound developed by Paddy Kingsland of the BBC *Radiophonic Workshop*. Using the facilities of the early twenty-first century, Maggs breathed new life into the idea, capturing the imagination of devotees as well as reaching a new generation of listeners, who were already cognisant with the gaming culture of larger-than-life visual effects and sound.

Such programmes enabled Maggs to refine ideas he had developed during the 1980s and 1990s, when he created sound equivalents of graphic novels, bringing familiar superheroes such as Batman, Superman and Judge Dredd to radio audiences on music stations, such as BBC Radio 1.

In the making of these cartoon-like 3-D productions, he evolved a genre that had existed almost since the start of radio, a form that transferred the vivid images and stories of comic-book heroes to a medium of pure sound, where prior knowledge of the visual characteristics of the protagonists informed the imagination of the listening ear. It was a tradition with a long and successful pedigree; in the USA, intermittently between 1932 and 1947, the CBS adventure serial, *Buck Rogers in the 25th Century*, based on the comic strip by Dick Calkins and Phil Nowlan, drew huge audiences, as did *Flash Gordon*, developed for radio during the mid-1930s from the comic-book character created by Alex Raymond. In July 1951, Radio Luxembourg broadcast the first episode of *The Adventures of Dan Dare, Pilot of the Future*, from the work of the Southport illustrator, Frank Hampson, familiar as graphic images through the children's comic paper, *The Eagle*. The series ran for five years. Roughly contemporary with this was the BBC Light Programme's *Journey into Space*, written by Charles Chilton. Although created exclusively for radio rather than as a development from other media, it accessed a largely young audience hungry for science fiction, while conveyed in a style that was less akin to the graphic comic than some of its antecedents and competitors. Nevertheless it created powerful visual images in the mind of its audience, running from 1953 to 1958, by which time it had been translated into 17 languages, and broadcast globally. *Journey into Space* is notable in the history of radio in that it was the last evening radio programme of its era to attract an audience larger than the one watching television at the time.

'In Space No One Can Hear You Scream'

Part of the significance of these productions, whether made as esoteric sound experiments, or for popular markets, was that they were literally 'unearthly' in the sense that a degree of separation existed between the idea of the human and the sounds the audience was hearing. Sonic innovation has long powered imaginative possibilities in both radio and film, and it is notable for its ability to unsettle the imagination of the listener. We are on unfamiliar ground, with no known point of reference, and apparently no human hand manipulating what we are hearing. We are alone in the haunted house, or on an alien planet, and something we do not understand is disturbing the ether. For example, the Theremin (originally known as 'The Etherphone', invented by Léon Theremin and patented by him in 1928,) was used to considerable effect in a number of science fiction films, notably *Forbidden Planet* (1956) in which Bebe and Louis Barron's score was the first in any genre to employ an exclusively electronic source. In it, the use of the Theremin blurred the lines between diegetic and non-diegetic sound, mirroring a plot that was groundbreaking in other ways. It was the first film to show humans journeying in a craft that moved faster than light, and also the first to be set from start to finish entirely on another planet. Thus, sound, image and story together immersed the audience in literally 'other-worldly' experience from the title sequence onwards.

Generating believable sound within the context of science fiction, supernatural or horror genres is the art of manipulating audience expectation, while avoiding the traps of cliché and predictability. The director and sound designer seek to challenge these in order to throw the onus back onto the audience's imagination, sometimes creating new, personal, hidden or alternative meanings in narratives and finding new ways to surprise and shock. There is a freedom of expression here too, because we are dealing with sounds, the authenticity of which cannot be verified. We may speculate on the sound of a dinosaur's roar, but we have no actual recorded evidence as a reference. The advent of electronic sound as in its use in *Forbidden Planet* enabled sounds made by machines—Moog synthesisers, Theremins, computers and so on—to underscore (as it were) meaning, setting and mood with sound that in itself was unfamiliar to the audience in every way, conveying a sense of futurity through its own strangeness. It is this very factor that makes the decisions taken by the director Stanley Kubrick in his groundbreaking film of 1968, *2001, A Space Odyssey* so remarkable. Here, Kubrick challenged what had become an accepted Hollywood model for the relationship between soundtrack score and image. In 1967, at the demand of MGM, he contacted the successful screen composer Alex North regarding music for the film. Kubrick had worked with North previously on his 1960 film, *Spartacus*. Yet it would seem that Kubrick already knew what he wanted his movie to sound like, and by apparently encouraging North to write for him, he was doing no more than appeasing the MGM management, members of which felt that the idea of using pre-existing traditional classical music would be a mistake.

In the event, the North score was never used, and with hindsight, the use of the music of Richard and Johann Strauss respectively is a key part of the film's meaning and identity. To audiences at the time, by wrong-footing their expectations of what science fiction should be, and radically eschewing clichés, Kubrick provided a whole new term of reference with which to interpret its elliptical meaning. *2001* begins with the famous theme from Richard Strauss's tone poem, *Also Sprach Zarathustra*, inspired by Friedrich Nietzsche's philosophical novel of the same name. In musical terms, this has been referred to as the World Riddle motif, a riddle that by the end of the music has not been resolved. We may perhaps interpret the unsolved riddle in Nietzsche's book and Strauss's tone poem as that of the universe itself—which thus provides an immediate statement and reference point for the beginning and ending of *2001*. Yet the meaning of novel, music and film is also independent of one another, although they may come from the same idea, as Michel Ciment pointed out: 'The death of God challenges man to rise above himself, and *2001* offers the same progression as in Nietzsche, from ape to man and from man to superman' (Ciment in Johnson, 137).

The long opening sequence of the film, set in prehistory, is for the most part unaccompanied by music. It is as Kubrick makes the astonishing time-leap from the killing fields of the apes to the graceful ballet of spaceships against a vision of the earth, taking the breath away through its transcendent visual beauty, that the famous 'Blue Danube' sequence enters, juxtaposing a familiar Viennese waltz, known across generations as a popular classic, with images of the future. Here, we might have expected something 'sci-fi' in the soundtrack to match the pictures; instead Kubrick indulges us—and himself—with a long dance in space. Yet there is meaning here too; as we discussed in a previous chapter, the Vienna that produced the elegance and romance of the glittering balls where Strauss's music was first heard was a city at the heart of the Hapsburg Empire, the might and wealth of which was built on power and domination. In the film, we have just witnessed early man learning how to kill, and here he is again, apparently civilised yet continuing to seek dominance, this time over the universe. (It is also worth noting that the waltz when it was new was considered as a risqué dance in some quarters, and there are undoubt-edly sexual metaphors to be found in the interplay between the slowly turning space station—so like a crinoline—and the approach and entry of the spaceship.)

The nearest we come in *2001* to music that anticipates the future is in Kubrick's use of György Ligeti's *Requiem for Soprano, Mezzo-Soprano, Two Mixed Choirs and Orchestra*, the sound of which supplies the leit-motif to accompany the appearance of the mysterious monolith, which 'reflects [the] idea that any technology far in advance of our own will be indistinguishable from magic and oddly enough, will have a certain irrational quality' (ibid., 136). This acknowledges a philosophical theme explored in the film's origin, the novel of the same name by Arthur C. Clarke. Ligeti's eerie voices weave and rise to a crescendo, and we note as we watch and listen, that the group of men gathered in the moon crater around the monolith share the same bewilderment as their ancestors in their first encounter with the shape, in ancient history. In among the care-fully placed meaning in the musical soundtrack, there are moments of profound silence. At some of the most dramatic moments, such as the death of Poole, when his air supply is severed, his vain fight for life ending as his body drifts away into infinity, we are *in* the void of space along-side him. Here, where another director might have demanded music that underlined the struggle, Kubrick gives us the horrific, impersonal anechoic stillness of a vacuum. In space, no one can hear you scream, and in real life, our tragedies do not come accompanied by soundtrack music.

Through silence, sound and music, there must always remain a thread of meaning, however elliptical, that links what we see with what we hear. It may be part of a puzzle in which meaning is either implicit or explicit, but however it is used, the sound is as much a part of the experience as the images. The editor and sound designer Walter Murch, from whom we heard earlier in the chapter, expressed it thus:

> Image and sound are linked together in a dance. And like some kinds of dance, they do not always have to be clasping each other around the waist…there are times when they must touch, there must be moments when they make some sort of contact, but then they can be off again…Out of the juxtaposition of what the sound is telling you and what the picture is telling

you, you (the audience) come up with a third idea which is composed of *both* picture and sound and resolves their superficial differences. The more dissimilar you can get between picture and sound, the more powerful the effect. (Murch in Weiss and Belton, 356)

Invisible worlds will always fascinate. Lodge, Raudive and Russolo sought documentary sonic evidence to demonstrate the paranormal, and fictional expressions of the same desire continue to attract the part of us that exploits the unknown in order to generate a kind of primal fear. There comes a point where the rational facts desert us, and we can only wonder and ask. Hamlet speaks of:

> …the dread of something after death,
> The undiscover'd country, from whose bourn
> No traveller returns, puzzles the will… (Act 3, scene 1)

When it comes to sound visions of the unknown, be it the distant future or the primeval past, there are really no wrong answers, unless they strike us as absurd: (A dinosaur mewing like a kitten would clearly not satisfy dramatic requirements.) For the most part, we can only speculate, and as long as the makers' speculation, whether scientist or storyteller, can convince the audience's imagination, the journey into the unknown, will continue.

Notes

1. Sullivan, Anita in *Radio Times*, 12–18 January 2019, p. 121. Sullivan's radio play, *Ropewalk House*, was broadcast first on BBC Radio 3 as part of the series, *Drama on 3*, at 7.30 p.m. on Sunday, 13 January 2019. It was directed at the Fish Factory Studios, London, by Joby Waldman of Reduced Listening Productions, with actors from Shunt Collective.
2. Liverpool Physical Society, Subscriptions Ledger, page 23 (Liverpool University Archives).
3. An artistic movement known as the 'Scapigliatura', the Italian equivalent of the French 'Boheme' (bohemian), was active up to about 1871, and included poets, writers, musicians, painters and sculptors. Members of the

scapigliati sought to radically reinvigorate Italian culture through influences from outside sources such as German Romanticism and the French bohemians.

4. Indeed, one of the early scenes in the drama takes place in 'the Princeton Observatory at Princeton where Carl Phillips, our commentator will interview Professor Richard Pierson, famous astronomer' (Cantril, 7).

5. The convention continues in the world of mobile Internet news 'apps'. In a time of rapidly changing news stories, we have become familiar again with interruptions, in the form of newsflashes from broadcasters such as CNN or BBC News alerting our devices to a new major event or developments in a continuing story.

7

Remembered Voices, Pictured Lives: The Sound of Recollection

Solitude

I am listening to a radio programme. It begins with about ten seconds of silence, and then, imperceptibly at first, a faint sound overlays it, or rather, emerges from the stillness: seawash, waves breaking on a beach. It grows in volume. It is a big sea, an Atlantic seawash, not a storm sea, but muscular, powerful, vocal and eloquent. I listen. The waves modulate across the stereo picture. I listen for another twenty-five seconds or so, and in my mind's eye I begin to see a shoreline, a small rocky cove with a shingle beach, an old broken boat hut at one end, at the other, low cliffs, with meagre vegetation sprouting from crags. The horizon is wide, looking eastwards, the sea, grey and flecked with white. It is a lonely place, and I am alone. Then, beside me, talking to me, a man's gentle voice, conversational, reflective, with a twang of a Canadian accent but something else too I cannot quite identify. The voice seems to emerge from the sound of the sea, just as the seawash came out of silence. It is like an imagining, a remembering, but it is coming through the speakers of my sound system from a recording. It is a kind of memory, but somehow different. I listen:

© The Author(s) 2019
S. Street, *The Sound inside the Silence*, Palgrave Studies in Sound,
https://doi.org/10.1007/978-981-13-8449-3_7

> This thing that has happened to you and me has happened to us over a period of a hundred years you know. The mad rush of this life has robbed us of the solitude, and really, the value of life comes from solitude…. (*The Latecomers*[1])

The sea sounds rise and fall, in counterpoint to the words, sometimes almost overwhelming them: some phrases are almost washed away, disappearing only to appear again a second or two later. It is almost like standing on the shore, watching a swimmer who occasionally vanishes from sight beneath the waves, before appearing again some way off. Above all it is a dialogue between two voices, one human and one elemental, although it could be said, both are tidal in their own way. I am listening to the opening of *The Latecomers, (Les Derniers Venus)*, a radio feature made by Glenn Gould for CBC Radio, first broadcast in November 1969. It forms part of a triptych of documentaries, collectively called *The Solitude Trilogy*, by Gould, who for many is most well known as a pianist, often controversial, but for me, always inspirational.

The mental picture conjured by these first sounds is a personal one, based on an actual memory. *The Latecomers* is about Newfoundland and its people, and this is a place I know; I have made radio programmes there and explored its history. The cove conjured here in my mind, with its boat hut and cliff, is somewhere I remember, near Cupid's Bay, where John Guy established the first permanent English colony in 1610 on behalf of Bristol's Society of Merchant Venturers. It is about an hour's drive from St. John's the capital of the island, and sometimes icebergs drift in from the north, sometimes there are whales. This is what I see on my mental screen: it is my picture, based on memory. Others will see other things, some will have more abstract responses. Above all, the interplay of these two voices teases and hypnotises like unfolding music in the mind. As the programme continues, more voices enter, and they overlap, merge and sometimes seem to contradict one another. It is confusing: I try to focus on one, but then another over-rides it, and I am drawn from one thought to another. It is rather as though I am in a room full of people, all talking to one another, and my attention is drawn first one way and then another, what is sometimes referred to as 'the cocktail-party effect'. There is some kind of a debate going on around me, but I cannot quite find a moment to pause and reflect on it. And all the time, there is the seawash; voices and waves rise and fall, volumes increase and diminish. It is radio, but it seems to disobey certain rules of production and balance.

I have known this programme for many years, and I had always assumed that the seawash was exclusively recorded around the coastline of Newfoundland; mostly, it would seem, it was. There is, however, so it transpires, somewhere within the programme, a 'fake' wave. After gathering his material, Gould spent many hours and days in CBC editing suites, cutting and preparing mixes of the voices and the sound of the water. It was crucial to his musician's ear that the balance between speech and seawash should be exactly right, because not only did the sea present a metaphor for isolation, but it was also a voice in itself. He later remembered, 'there was one scene for which I just couldn't find the right surf sound; I tried everything, every tape of sound we'd recorded in Newfoundland, and nothing quite matched the mood of the voices in that particular scene' (Gould, quoted by Friedrich, 196). Mentioning the problem to associates at the CBC, he was presented with an alternative by a sound crew who had just returned from location recording for a programme about Charles Darwin's voyage on *H. M. S. Beagle* off the coast of Ecuador. Included in their recordings were those of wave sounds; Gould listened and eventually found one that rhymed with the sense, timbre and rhythm of the Newfoundland voice. 'So we ended up borrowing a tape from someone who had just returned from the Galapagos, and that worked perfectly' (ibid.). On the face of it, one wave may sound much the same as another. Yet Gould's biographer, Otto Friedrich makes the point that 'all the traditions of history, scholarship, and even journalism dictate that every statement must at least try to be true, that the thing should be what it pretends to be…To "match the mood" of a Newfoundlander's voice with the sound of a wave breaking on the Galapagos Islands is almost a contradiction in terms, an absurdity' (ibid.). Thus, does it become a question of culture? On the other hand, without this information, who would know, and even *with* the knowledge, who can tell where that moment is in the programme? Only the makers could ever say. In the end, the answer may lie in a debate between temperaments; a journalistic documentary may seek truth and answers, while a poem will also seek truth, but through questions and images. An authentic human voice belongs to its place. It would seem that the language of seawash is international; but is it?[2]

This book is about sound and the imagination; we might argue that the end justifies the means, that what matters is the sense, meaning and pictures evoked in the mind, rather than the process of placing them there. Gould's programme is a lyrical statement made of sound, and it contains an essence; he is using the *musique concrète* of a place to generate an overall poetic thesis. Gould wrote of *The Latecomers*, and the process of making it:

> The Newfoundlander is, first of all, a poet. Spirits of celtic bards linger on among these people; a sense of cadence, of rhythmic poise, makes their speech a tape editor's delight. Even when caught out with little to say – and that happens rarely – they say it in elegant metrics. But mingling with the urge to turn all observations into verse, a blunt saga-like dispatch of detail gives point and purpose to their story. Perhaps the fact of life against the elements – as for the Iceland and Greenland peoples – disciplines their stanzas, gives an underpinning of reality to their ever-ready impulse to fantasize. (Gould, programme note[3])

Newfoundland's east coast is populated by families whose origins are from the west of England, from Dorset, Devon and Cornwall, while the south coast of the island is made up of the descendants of Irish immigrants. I hear my own kinfolk in these voices, and I remember going out in a small boat from just such a cove as Cupid's with a fisherman, catching squid, with a number of other men, like a small convoy off the rugged shoreline of Random Sound. I remember the indeterminate tang of his accent as he called across the water to another nearby fisherman: 'I got a feller here from the 'omeland. I told 'un, 'e's got a long way to swim to go 'ome!' I will probably never go 'squid-jigging' as they call it, with him again, but I now suddenly think of him again, through a chain of reactions set up by the sound of a human voice and the sea at the beginning of Glenn Gould's 1969 radio feature.

The Latecomers is the second in *The Solitude Trilogy*; the first being *The Idea of North*, in which a quintet of voices merge and emerge from the sound of a train journey. We are in a railway dining-car, listening to, in the words of Gould's technical supervisor, Lorne Tulk, 'a tapestry of the

Canadian experience above the 60th parallel. It represented both the real and imagined effects of geographical isolation, and it was the jumping-off point for Glenn's exploration of solitude via radio documentary' (Tulk, ibid.). *The Idea of North* set a model in Gould's inner alchemy that would inform the programmes that followed, and underlying everything is the idea of speech and sound as music, a multi-layered structure that evolved to its most contrapuntal and yet in some ways its clearest and perhaps most perfect form in the third and last programme in the series, *The Quiet of the Land*, broadcast by the CBC in March 1977.

Coming to understand better the thinking behind Gould's fascination for radio, I realise now that I cannot divorce that thinking from the ideas behind his great recordings of Bach, because for Gould this *is* music. In a 1968 interview, he said:

> What we've tried to do…is to create what I've grown rather fond of calling "contrapuntal radio," which is a term I've picked up from a fondness of contrapuntal music, and tried, rather arbitrarily, to attach to another medium, where it has not belonged in the past… Some of our most aware experiences are gleaned from sitting in subways, in dining-cars on trains, in hotel lobbies – simultaneously listening to several conversations, switching our point of view from one to another – picking out strands that fascinate us. (Gould, quoted by Payzant, 131)

The speakers in these programmes were recorded separately, and yet Gould makes them sound as though they are conversing together, in some cases actually arguing. Understanding this, my concern about the fake wave gains a new perception; this is a documentary drama, more, it is a theme, fugue and variations. Gould is not concerned about the authenticity of the wave, but its musical correctness in the context of the overall musical logic in his programme. He blends sounds based on their *musical* relationship to one another, raising and lowering volumes from *ppp* to *fff* according to the 'score' that is directing him inside his head. It is artifice, a constructed work. If it touches emotional or mnemonic chords, it does so through the same process (Gould hopes) that would affect us if we were listening to an evocative piece of music. For the last part of the trilogy, *The Quiet of the Land*, Gould recorded his material on location with a Mennonite Community of Manitoba, in a feature that explored how members of a religious group, long separated from the mainstream of modern Canadian life, coped with the pressures and strains of the twentieth century as it roared and shouted around them.

To someone familiar with more traditional documentary journalistic production techniques and structures, Gould's radio work can provide a challenge. These overlapping themes, phrases and shades of meaning, sometimes one fading into another, or a voice coming in over another from a different direction, or with a divergent idea, clashing and obscuring an emergent meaning, can be frustrating. Even to the tuned ear, the experience can be frustrating. As the radio documentary and feature producer, Piers Plowright, has said, 'I just long for silence, for a moment of space in which the still small voice of humanity can be heard. It's one of the Gouldian paradoxes, that a man who could use silence so beautifully in his playing, gives us so little of it in his radio documentaries. And surely, silence is one thing that Northern Canada is full of' (Plowright, 2007[4]).

The Gould radio documentaries—including programmes he made about some of his musical heroes, such as Arnold Schoenberg, Pablo Casals, Leopold Stokowski and Richard Strauss, all produced between 1962 and 1979—demonstrate the nature of his experiment in sound, to stretch the listener's ears and mind with multiple, simultaneous strands of information, woven together in contrapuntal textures. In *The Quiet of the Land*, the interplay forms its most subtle montage, blending layers of ambient sound, music and speech in the form of voices that overlap one another as members of the community share their thoughts about the Mennonite idea of separation from the world and the intrusion of modern society upon their lives. In this last programme, Gould has his forces most fully under control, managing as Piers Plowright says, 'to get the balance between sound, word, and music, exactly right. In this most explicitly religious of his programmes, paradoxically, Gould the preacher takes a back seat and allows the people and place to speak for themselves. There's no relentless ground bass, very little "cascading" of voices, music is often subtly held back as the people tell their stories' (ibid.).

To be fully understood in terms of their underlying philosophy of sound, these programmes are best listened to in the same frame of mind with which one might approach, for example, his recordings of Bach's *Goldberg*

Variations. He made two recordings of this great masterpiece: in 1955 and 1981. Between the two, he increasingly eschewed the medium of the concert hall and live performance and embraced the studio as a means of seeking creative perfection, meticulously piecing the final recording of the work's 32 elements together from numerous takes, into a jigsaw which he considered to be finally a definitive statement. The question here is, does a 'live' concert performance, with ambience, atmosphere, acoustic (and the occasional cough or shuffle of feet) make for a more powerful imaginative experience? Is The Beatles' *Sergeant Pepper's Lonely Hearts Club Band* any better or worse for having been 'constructed' in a studio than if it was a souvenir of a 'live' performance?

Glenn Gould's radio work often requires intense concentration in order to engage with it, and each of the *Solitude* features lasts about an hour. Nevertheless, the best of them can yield considerable rewards. The over-all picture conveys a unified meditation, but above all it is the sound of the voices, the principal musical instruments in these complex compositions, that insinuates itself into the memory. The cumulative meaning behind the themes upon which they expand, develop and with which they interact with the mind are indeed like a fugue: their timbre, pitch, tone and modulation, full of subtle mood and meaning that creates a picture framed within the form of a radio programme, hung on the wall of a programme schedule. The programmes in *The Solitude Trilogy* may perplex—or indeed irritate—some listeners at first hearing. Yet, long after they end, somehow something in them continues to haunt the mind, be it the rumble of trains, the whisper of seawash or the tidal murmuring of voices. While they are not quite singable from memory, they do have something of the quality of a remembered melody at times, if they are given the opportunity to sink into the bed of the subconscious, to be played back through memory. Perhaps too often we expect the response to a creative work to be instantaneous; in some instances, allowing it to live longer within the mind can return richer rewards.

People, Remembered and Recorded

It is in such ways that the pictorial nature of the sonic imagination enables us to conjure images of people and places we have known, as well as our capacity to evoke imagined pictures of those we have never met or seen. Personal memory can be highly potent, and this is particularly true of the human voice; poignant too, because unlike the objects with which we surround ourselves, it is invisible. Informal family recordings offer us the person we knew. They are the sonic equivalent of a street photograph of a known and loved place, special for the very reason that they provide us with the memory of a day and a place when life was normal, even ordinary, and deeply touching in recollection for that very reason. The asides, the heard smile that reveals the person, even the stumbles and unedited hesitations, all these give us our loved ones back to us in a way to which written stories and recollections themselves can only partly contribute: the latter is the documentary, the former is the human event. The power of the voice to recall a personality is so strong that, while we may cherish our recordings of loved ones, we may sometimes find it difficult to confront their actual sound: it is a preserved part of the physical person in a way that goes beyond the visual remembrance of a snapshot in a family album. A photograph is a moment in time, a split second of existence. The sense of the place or the person immediately before or after its taking is lost to us. It may be an aide-memoire, but it can seldom compare to the experience of listening to sound for a sense of the living human. This is because a recording is temporal; it has to be experienced in terms of time. As we passed through time when we made it, so we pass through time again as we listen. The recording of the voice to which we are listening, even if it is the voice of a dead person, moves along on this temporal journey, so for the time it plays, we have a feeling that the person lives again, and we regain their presence. We may dwell on the written word, sense the sound in it, but the spoken word was there first. There was human sound even before there were human marks on cave walls. The sentence on the page is actually a representation of real language, which is sound; it is the best we can do, and as we have seen, the written word can convey simulations of speech sound vividly. Yet, as Walter Ong emphasised in his book, *The Presence of the Word*, the former grew out of the latter: 'Writing is a derivative of speech, not vice versa', and today 'we habitually mingle speech and writing so much that when we discuss language we hardly know whether to refer to oral performance or to written work or both' (Ong, 19). Homer, for example, exists as text and is thus preserved for our time, but it is now, as it was then, meant first to be heard. The origins of poetry are spoken:

> We tend, for example, to think of early oral cultures before the invention of script as simply illiterate or pre-literate, that is as cultures without writing or before writing…For hundreds of years such cultures had existed without anyone's thinking even of the possibility of script. To think of them in terms of their relationship to script is the equivalent of working out the biology of a horse in terms of what goes on in an automobile factory. (ibid.)

It is our capacity to be identified through sound that is one of the most potent things we possess as an individual. We know one another by sound as much as anything. We all are animal in nature and find our way through a map of sound memory deeply embedded in instinct and evolution. Likewise we are both transmitters and receivers, possessing the essential ability to send, receive and—importantly—understand signals. This takes us beyond the ears, just as it takes us beyond the larynx. As we have seen, we sound ourselves with our whole body through vibration in the upper, middle and lower zones. The voice is much more than what happens in the larynx; the sound originating there is transformed by a potentially huge pallet of timbres from throughout the head and trunk in particular. Likewise, with our listening, we hear one another through vibration, acoustics and circumstances (according to location and mood) among other factors. Our memory of a familiar voice is made up of interpretation based on individual shape and form; even in the case of profound deafness, it is crucial to the cartography of the human spirit as it encounters and negotiates the world around us.

The ability to record saves the aural evidence of lives lived, while for the programme maker, editing (of tape, and subsequently through digital technology) retains control and enables shaping and often the steering of content towards an editorial agenda. In the UK, producers such as Charles Parker argued that by editing, the broadcaster aided meaning through releasing the thought from a cage of speech patterns that sometimes concealed or blurred the sense intended by the speaker. Parker claimed that in so doing he was setting the thought like a jewel in a ring, framed by music or complementary sound effects, and thus giving it its full power. Parker, with his frequent collaborators, Ewan MacColl and Peggy Seeger, could be argued to have made 'composed radio' like Glenn Gould; yet for Charles Parker, the social philosophy guided the content, and the framing of the word, the phrase or the sentence was never concealed, but rather enhanced by the sounds and music around it. Common between both producers was the sense of craft, the opportunities offered by the editing pencil and the razor blade to shape the spoken word and other sounds into the most powerful structure in support of an overall programme. The voices in works such as *The Radio Ballads*, a series of eight features broadcast by the BBC between 1958 and 1964, are those of real people recording in real places, distilled and honed to what Parker considered to be the essence of the statement that was in their mind as they spoke. Sometimes, we may find it hard to put our thoughts in order. Far from using the voice simply as a sound or musical effect, Charles Parker was an enabler who aided and refined the process towards perfect expression.

Yet for some, such as Roland Barthes, something is still lost in this process; the stumbles, hesitations and repetitions of vernacular speech often show us the personality, and more, the emotion behind the words. Phrases such as, 'Do you know what I mean?', 'Do you understand me?' and other apparently incidental words and expressions 'are yet in some way discreetly dramatic: they are appeals, modulations – should I say, thinking of birds' songs? – through which a body seeks another body. It is this song, gauche, flat, ridiculous when written down – which is extinguishing our writing' (Barthes 1985, 5). To this we might add that it is a regret when, although considered necessary in some journalistic contexts, the struggle towards a meaning is edited out in a broadcast, leaving us with the end product of a sound-bite. One of the curses of modern broadcast news journalism must surely be: 'I'm sorry, but we're running out of time'. Just as an internet search can provide us with an answer without the process of enquiry, so the reduction of speech to a unit of information may be seen as a dehumanising of vernacular language. For oral historians, the voice and the turn of phrase matter deeply. The British Library's Oral History Department and its equivalent at the US Library of Congress are treasure troves of recordings that bear witness to not only what happened, but how it was heard and how the voice makes music from the event. A former slave, Anna Woods, was interviewed by the Library of Congress about her recollection of the moment she became a free person. Hers is an authentic, audible voice, right down to the seemingly incidental and apparently insignificant recalled details, and it speaks aloud even when transcribed from sound to paper:

> We wasn't there in Texas long when the soldiers marched into tell us we was free. Seems to me like it was on a Monday morning when they come in. Yes, it was a Monday. They went out to the field and told them they was free. Marched them out of the fields. They come a-shouting. I remembers one woman. She jumped on a barrel and she shouted. She jumped off and she shouted. She jumped back on again and shouted some more. She kept that up for a long time, just jumping on a barrel and back off again. (Library of Congress, quoted in Lester, 136)

It is in the nature of true oral history that we touch the tap root, because it was the first way in which history was recorded, through personal memory. The oral historian, talking with people about their recollections, may encounter anomalies, because human beings are not academic documents or manuals: 'They find that people whom they interview do not fit easily into the social types presented by the preliminary reading…they encounter the problems of bias, contradiction, and interpretation in evidence. Above all, they are brought back from the grand matters of written history to the awkwardly individual lives which are its basis' (Thompson, 12). Our experiences of life have been carried within us since they first occurred and that may be perhaps for a lifetime; in the intervening years, memory, imagination, new knowledge, information and prejudice may change and colour the facts as they once happened; what remains, however, is what Roland Barthes called 'the grain of the voice', and it is seductive in its persuasion. The phrase, 'to capture the imagination', has no truer meaning than this, and the birth of portable recording technologies during the twentieth century enabled the preservation of oral moments that safeguard voices and their witness. Perhaps few oral historians typify this impetus and motive more than the Welsh-born teacher, writer and folklorist, George Ewart Evans. (1909–1988) who became a dedicated and humane collector of oral history, (or 'spoken history' as he preferred to call it) and the oral tradition in the countryside of East Anglia in England from the 1950s to the 1970s. He produced his books from material collected at first hand, and it is in these that the voices of the past can be heard explaining the customs and ways of life that were fast disappearing through their own turns of phrase and dialect. In Chapter 3, we discussed some of these voices, and their communication of the preserved witness of lost times. During Evans's childhood, he noted at firsthand the struggles of a Welsh mining community during the years of the British Depression of the 1930s. He observed the affects of mining disasters, strikes and most poignantly the loss of his father's shop and the experience of his resultant bankruptcy. After learning to operate wireless equipment in the Royal Air Force during the Second World War, he moved to the village of Blaxhall in Suffolk in 1947 and began writing scripts for the BBC. It was at this time that he wrote his first book, *Ask the Fellows Who Cut the Hay*, which focused on the people he met around him; it was eventually published in 1956, beginning a career as an author that lasted until his death. Writing some thirty books later, in *Spoken History*, published in 1987, a year before his death at the age of 79, he reflected on his arrival in Blaxhall, and how the human sounds he heard around him, came as a revelation:

> One of the first things that struck me was the speech of the natives of the village. It alerted me to the age of the community we had come to and gradually persuaded me that this dialect or *variety* of speech, as some scholars prefer to call it, avoiding the constricting associations of the word "dialect" itself, was suitable as a vehicle for transferring the history of the East Anglian people to a new synthesis. As well as being signally appropriate to the description of East Anglian farming, it bears the accent of an older language and a much healthier one than current English, its unadorned purity being unpolluted by the numerous abstractions that have crept into our everyday discourse. (Evans, xv)

Reading the works of George Ewart Evans, we come to realise that the relationship between language and identity is of a profound depth that can hardly be exaggerated. By capturing these voices before the homogeneity of a spreading 'classless' and urban speech diluted and blurred them, he proved that the sound of authentic place-born language carries the characteristics of that place into the imagination just as much as does the visual memory of a particular field, hill, building or street; in short, these voices are (or rather were) sonic events with the same reality as the visual events of the terrain from which they grew. The two are interchangeable, and in the mind, the sound of one contains the capacity and power to vividly evoke the image of the other, both as an historical entity and as a present vestige. It is held in the very 'grain' of the voice, just as Anna Woods's language in recollecting the moment of being freed from slavery in Texas preserved more than history through the very syntax of her utterance. For Evans in post-war East Anglia, it was this quality in everyday speech from the villagers around him that mattered almost as much as the content of what they had to say; 'sometimes even displacing my intentness on their description or story to the actual manner of its telling…Sometimes…you hear gems in casual conversation, such as a young boy's phrase, undoubtedly heard from his grandfather when he was describing a heavy snowfall: "It snew a masterpiece"' (Evans, 162). In typing that last wonderful phrase, I have been confronted by an objection silently expressed by my computer; I contradict it and type on: 'Snew' *is* an authentic word. We understand that language evolves, and in Chapter 9, we shall revisit the subject, exploring how paradoxically a new authenticity emerges through speech that grows more universal and louder as it becomes less vocalised and place-centred. The preservation and acknowledgement of language and speech as part of changing and disappearing cultural (and physical) landscapes are thus thrown into relief as being of the utmost importance and value in mapping the inner sonic imagination. Robert Macfarlane's beautiful book, *Landmarks*, seeks to do just this: to hold language within the context of the regions and smaller localities of which it was a part, a crucial aspect of identity that can be a sonic badge worn with enormous pride to whoever it belongs. At the same time, such vivid evidence of origin can prove to be a source of varying degrees of suspicion, prejudice and bewilderment for some who feel distanced by class or location from their origins, and thus, in some cases, threatened by differences. Macfarlane focuses on geography, highlighting words and turns of phrase that remain in local language long after the social and industrial conditions that spawned them have changed. For example:

> …The miners working the Great Northern Coalfield in England's north-east developed a dialect known as "Pitmatical" or "yakka", so dense it proved incomprehensible to Victorian parliamentary commissioners seeking to improve conditions in the mines in the 1840s. The name "Pitmatical" was originally chosen to echo "mathematical", and thereby emphasise the craft and skilful precision of the colliers. Such super-specific argots are born of lives lived long – and laboured hard – on land and sea. The terms they contain allow us glimpses through other eyes, permit brief access to distant habits of perception. The poet Norman MacCaig commended the "seagull voice" of his Aunt Julia, who lived her long life on the Isle of Harris, so embedded in her terrain that she came to think *with* and speak *in* its creatures and climate. (Macfarlane, 5)

While a word such as 'Pitmatical' speaks of a consciously constructed requirement to create a precise term for an activity, much vernacular speech is the result of evolution, either of dialect or, beyond that, of incoming or native languages, elements of which have become absorbed as a shoreline retains seashells. Macfarlane delights in inviting us to roll these sounds around on our tongue: 'Drindle', a diminutive run of water (East Anglia), 'aggy-jaggers', mist that forms along the sea edge (North Kent Coast), 'puxy', meaning deep mud (Exmoor), or a 'roddie', a small footpath, as heard in Caithness. All these words belong to the English language; by simply reciting a few of them we come to realise how impoverished modern speech has become. If it is not actually a duty to widen our vocabulary to absorb such words, we would surely find it enriching and enhancing to do so. Even in the common usage of today, we find words the meaning of which may be undermined by local ownership; for example, the word 'shard' might be understood to mean a sharp-edged fragment of broken glass, metal or ceramic, (or to a more current generation, an architectural feature on the London skyline). Yet Macfarlane reminds us that in the southern English county of Wiltshire it has been used to identify a gap in a hedge (ibid., 373). Even more startling to modern consciousness might be the realisation that the word, 'broadband', has another meaning far removed from its ubiquitous modern appropriation; this, Macfarlane tells us, has been a term employed within certain rural communities of Northern Ireland as part of the vocabulary surrounding the practice of 'opening and spreading out beets of flax in order to dry them' (ibid.,

353). It would seem that we abandon some words and adopt others; even within recent cultural history, the word 'wireless' evokes two forms of media dissemination. Largely obsolete now is the old term relating to radio transmission, while ironically, the very technology for which it has been appropriated today is one of the reasons behind the decline of dialect and verbal communication.

The concept from very early in radio history of what was to become known as 'the BBC voice' spread exponentially, as did the idea of a vocabulary that must be instantly understood by as wide a range of the populace as possible. Phrases such as 'received pronunciation', 'standard English'—or worse, 'King's English' or 'Queen's English'—speak of a denial of individuality through a form of linguistic cleansing that is more part of modern culture than history. The actor Ian McKellen grew up in Lancashire during the 1940s:

> As a child, I spoke with a flat northern voice – two voices actually – a broad, aggressive accent for school and the playground and a milder, more yielding tone at home. My would-be posher schoolmates went to "Elocution" to have their native sounds taken out like tonsils. Similar violations were perpetrated in drama schools at the time. At Cambridge, undergraduates from public school mocked my northern vowels and I taught myself to disguise them. (McKellen in Rodenburg, vii–viii)

Dialect and accent spoke so eloquently to prejudicial ideas of place that it was shorthand for 'lower' or 'working' class. Times change, and today, actors and public speakers such as McKellen are in the privileged position of being able to champion dialect on radio, stage and screen. Yet it still remains that the keyword is 'privileged'. The speech expert and teacher, Patsy Rodenburg, makes the point that an often subconscious prejudice is inherited in many of the activities society has come to class as 'high art' and culture: 'As soon as we open our mouths and speak we are judged. Instant assumptions are made about us by others; about our intelligence, our background, our class, race, our education, abilities and ultimately our power. As listeners we do this to each other all the time' (ibid., 4). Such uninformed sonic prejudice is a darker, negative side of imaginative consciousness; we document from minute to minute, and as living archives, we add to our library of impressions. To reverse a social process that has evolved for decades, even centuries, may be beyond the bounds of realistic possibility, but an awareness of the diversity of speech and language is in the end no more than human geography; by listening and considering origins, we enrich and inform our very selves. It is, as so often, the poets who can lead the way; writers such as Norman MacCaig, David Jones and Kevin Crossley-Holland have encouraged us through their work to become inquisitive as to the stories behind words and phrases. Crossley-Holland in particular, as a scholar and translator of ancient Anglo-Saxon tales both for children and adults, sometimes challenges the reader with words that might seem unfamiliar to modern ears, yet once explored, open up unexpected nuances of meaning linked to terrain. Thus, in the very title of his 1986 collection *Waterslain*—'an old Norfolk word meaning "flooded"'—he generates a landscape both real and imaginative that takes its identity from 'a village on the north Norfolk coast…The cycle of poems bearing its name describes its inhabitants and concerns,…[remembered] from childhood'.[5] Here is the bridge between the ear that observes and analyses, and a sense of intimate cultural identification. We are all from somewhere and from a past; we originate in the lives and terrain of our ancestors, whether we seek to celebrate it or deny it.

Listening to voices as documents, either social or historical, however authentic, is one thing; we may be engaged, moved, shocked or inspired, but we remain, to one degree or another, objective in our appraisal of what we are hearing. This changes when we are given the opportunity of listening to the voice of a lost loved one or even ourselves. In both instances, a recording, however perfect, removes layers that are both technical and imaginative, while it replaces some of those layers with subjective reactions that are deeply personal. The essayist, Hazlitt, said 'the sudden hearing of a well-known voice has something in it more affecting and striking than the sudden meeting with a face: perhaps, indeed, this may be because we have a more familiar remembrance of the one than the other, and the voice takes us more by surprise on that account' (Hazlitt, 125–6). The written words provide the information, but a physical sound, reveals their full meaning and often offers an alternative interpretation. Making a call to home from a distant location, a much known and loved voice answers the telephone. 'How are things?' we ask. 'Everything's fine' comes the reply.

We have known this voice for a long time, perhaps even for a lifetime, and we *know* in an instant how true or not is the statement contained in those two words. If this intimate knowledge is based on experience, how much more intimately do we know—or think we know—our own personal sound? Trevor Cox, a scholar of acoustic engineering, gives us a salutary reminder: 'We spend a lifetime listening to a voice that appears more boomy than the one others hear, because bone vibrations carry the sound internally from the larynx to the ear and boost the bass. A recording quickly reveals that the vocal identity we present to others does not match our own inner voice' (Cox, 2). As we have seen, the voice is our mind's bodily representative, and we shall discuss further in the final chapter of this book. It is our most intensely personal means of expression as both a conduit for communication and as a source of identity. It is, we might suggest, one the most familiar sounds we know, without even recognising the fact. Yet how many people, hearing their own voice played back to them for the first time, respond with the words, 'I don't sound like that!' In saying this, they are not telling a variant of truth, only of perception. Thus, the fact is that imagination lies at our very core; based on what we hear vibrating inside ourselves, we *imagine* the oral effect we are having on those around us.

Listening to the recorded voice of a departed relative or close friend can be, as we have said, even more difficult to come to terms with. I can imaginatively *hear* the voices of my mother and father, who died in 1985 and 1987, respectively; my mother's County Monaghan dialect stayed with her all her life, through nearly fifty years of exile from her native Ireland. I could only 'hear' it, however, when we spoke on the telephone, partly because the technology removed certain frequencies, exposing the essence of her accent, and partly because, with no visual prompts, I was listening to the sound of her with a degree of objectivity, distanced from her presence. The same happens with much more emotional impact now, when I listen to an actual recording of her speaking conversationally. In the meantime, I can 'hear' her speaking, imaginatively, even more perhaps than I can imagine her face without the aid of a photograph. I also have recordings of my father's voice; I can 'hear' him by listening in a similar way, although now also through a kind of qualifying filter that grants a certain amount of objectivity. William Street was a naval officer before and during the Second World War, and a career in the British navy to some degree smoothed out his Hampshire accent, absorbed during a childhood spent with his rural family on farms deep in the heart of the countryside. Yet listening to the voice played back, I hear that dialect so much stronger than I thought I remembered. The producer, Clare Jenkin, was inspired to make a documentary for BBC Radio 4, after experiencing her late father's voice, played back on tape. 'I'd forgotten until I heard his voice on the tape, how strong my father's Scottish accent was. I knew he was Scottish of course, but I'd forgotten how Scottish he *sounded*…We know that people will keep answering machine messages, to hold on to the voice. It's part of the person, like a lock of hair' (Jenkin in Street 2015, 79). Sometimes, imagination and memory, however apparently faithful, are simply not enough in our desire to preserve what is no longer physically there. Then, as if emerging from the seawash in a Glenn Gould radio documentary, a personal memory, a fragment of a voice, sharply focused through the overlapping murmur of voices that gradually merge through time, touches the heart. We have already considered the significance of the sound of a place, and the lost whispers of history; we can now add to that, the sonic presence within ourselves, the remembered land we carry with us, and which makes us who we are.

Notes

1. Gould, Glenn. *The Latecomers* (November 1969) from *The Solitude Trilogy*, November 1969, *CD 1*, 00.38–1.32. Toronto: CBC Records, PSCD2003-3, 1992.
2. This relates to our earlier discussion of sound as an imaginative event and the question of cerebral, cultural and emotional authenticity. In the view of Cheryl Tipp, Curator, Wildlife and Environmental Sounds at the British Library: 'From a purely scientific stance, I don't think that one shoreline is capable of producing the same sounds as another. Substrate type, slope, vegetation, geographical position etc. all influence how the shoreline sounds. Now whether we, as humans, can notice these subtle differences is another thing entirely. It's easy to tell the difference between a shingle beach and a sandy shoreline, but if two stretches of coast are similar then the ability to tell them apart becomes increasingly harder. And

of course, the sound of a particular shoreline isn't fixed. That too changes from day to day. Weather, the moon, coastal erosion etc. all play their part' (Tipp, Cheryl, personal communication, 31 January 2019. Reproduced by permission).

3. Gould, Glenn. *The Latecomers*, Insert sleeve note, CBC Records, November 1969. Toronto: CBC Records PSCD 2003-3, 1992

4. Plowright, Piers. 'Muzak, Music and Monologues: The Mind of Glenn Gould as Revealed in His Radio Documentaries'. Unpublished lecture delivered as part of the Canadian Museum of Civilisation's *Glenn Gould: The Sounds of Genius* Symposium, in Gatineau-Ottawa, 29 September 2007. Reproduced by permission.

5. Crossley-Holland, Kevin. Cover Note to *Waterslain*. London: Hutchinson, 1986.

8

The Singing of Statues: How Art Sounds

Dialogues with Stone

In his remarkable short story, *Schwarz*, Russell Hoban's narrator visits the hall of Oriental Antiquities in the British Museum. He stops in front of a black stone lion, from the Sung Dynasty, dated AD 960–1279:

> I had my Walkman in my shoulder bag, in it a cassette of the Thelonious Monk quartet on its 1961 European tour…The tape was at the end of 'Round About Midnight' and just about to begin 'Blue Monk.' I put the headphones on the lion's head and started the music. I had my left hand on the lion's head, and when I felt the stone begin to hum I took off the headphones. The lion blurred, the floor shook a little and several people, noticing perhaps some disturbance in the air, turned to look as we walked away. On the plinth there was still a lion that could be seen and touched but the stone was empty. (Hoban, 33)

It is an eloquent, if surreal expression of human and artefact interacting, and the two go on to explore shared ideas and feelings in a strange journey in which everyday London becomes almost ghost-like against this alternative reality. We affect a space by our very presence in it, even as we seek

© The Author(s) 2019
S. Street, *The Sound inside the Silence*, Palgrave Studies in Sound,
https://doi.org/10.1007/978-981-13-8449-3_8

to record the experience of that space. A room full of listeners changes the character of that room. A single person recording an unpopulated landscape negates the idea of human absence. By being present to experience, by being a witness to the visible and audible, our consciousness contributes actively to that experience. We are makers at the same time as we are observers. Likewise, as in *Schwarz,* an apparently mute image does not offer silence but a receptacle in which the imagination can create its own personalised sound world, just as in darkness the mind may generate its own pictures to people the void. A sculpture, a building or a text can call forth a mental response from us beyond the visual; it may be conscious or subliminal, an internally articulated thought, a sonic signal of interactive emotion. We may have linguistic limitations, but to move beyond words is to enter a world of sound suggested by the other senses that truly knows no frontiers. The coming of speech to motion pictures immediately placed walls of linguistic limitation that had not existed before. We are participants in a sound work, and we personalise the world internally, but it is an acquired skill, to train eye and ear to relate to one another, utilising memory to provide references.

It is through hearing that we 'see' imaginatively, and in this respect, instrumental music is truly international. Such powers of radio and the sonic arts have long been recognised. As the old truism has it, 'in radio, the pictures are better'. I would like however to explore here the possibility that this imaginative sonic signalling between image and ourselves is in fact somewhat more complex, and that we possess the capacity to make imagined sound from pictures, just as we have the ability to make pictures in the mind from sound. The sonic expression suggested to our imagination by a work of art may be a literal interpretation evinced by clues uttered by the object itself, or it may manifest itself as an abstract or associative sound, based on memory, a form of sonic feeling or a particular theme or piece of music.

The novelist, Jeanette Winterson, has written, 'Art looks like an object but it isn't one' (Winterson in Gormley, Richardson and Winterson, 112). In fact, it is an event and an experience. More, it is a voice both of its own as 'heard' by the viewer, and in its origins at the hands of its maker. It is above all else, a conversation and a relationship, a meeting and an immediate communication. In my previous Palgrave book, I explored a little of this idea, but here I wish to take it further, in a contemplation of two- and three-dimensional works of art and their ability to represent frozen sonic moments. The idea of sound contained within stone has its roots in the classical world and the relationship between the visual arts and poetry. Beyond this, the physical presence of an object evokes imaginative sound, even at its most subtle, as a kind of vibration of the emotions and intellect. While the world of antiquity saw fewer boundaries than in our own time between science and art, and indeed between the arts themselves, lines of demarcation were drawn in the late eighteenth and nineteenth centuries. Words are only one form of sound. An object, be it a statue or a building, may possess its own presence as a poetic form that bypasses the word. It may be a sound like a scream or a shout, or indeed it may actually contain profound stillness, and this returns us to the idea of silence as its own sound. The imaginative sound of an artwork is mitigated by the contextual sound of the place in which it is 'heard'. So a statue of Samson slaying Philistines, standing as it does in a public gallery within London's Victoria and Albert Museum may evoke the noise of battle and death in the viewer's mind, yet around them remains the chatter, laughter and general hubbub of the place itself. The object may for a while transport us into its sound world, so that we forget this context, but it is our own world to which we return as we move on, while the killing fields remain behind, frozen forever in the act of violence. This makes the scene somehow more shocking. Christopher Logue's poem, 'The Flaying of Marcyas', depicts an act of unspeakable horror, echoing the statue of the same name. But the words, like all sound, move through time, while Marcyas stays fixed in his agony.

We 'hear' Antony Gormley's statues of human beings against the backing track of the landscape, the sound of which constantly evolves, while the figures seem to observe the terrain, lost in their own thoughts. They may shock, or at least startle us, at first sight, these apparent aliens who are reflections of ourselves, yet when we approach them, they seem to possess a poignant voice that enters us by other means than the auricular. A Gormley figure on the shoreline can seem enigmatic, as Winterson says: 'we do not know who he is or what he has seen. We cannot read his thoughts. But we can read our own. The iron body is a magnet for thinking' (ibid., 36).

We have heard of legendary statues that have become empowered with the ability to walk, weep and speak. Imaginatively, every statue that engages us fully seems uncannily to possess the potential to do so. Placed within the landscape, with all the elements singing around it, this potential seems only a whisper away. Our relationship with a work of art is as intimate or as detached as we make it. Winterson's thinking seems to agree with Hoban: 'Anyone can own a work of art. The moment you recognise it, really recognise it, both its significance and its significance to you – its image hangs in a gallery or remains on the street, but it locates inside you' (ibid., 113).

Public art is powerful in this respect simply because one comes upon it rather like a rock or a cleft in the landscape. It is *there* as we are, on equal terms, and it is invested with a voice in the same way that everything that surrounds it interacts with us as we pass through: 'Public monuments are semaphore systems made out of solids' (ibid., 10). Certain images are profound expressions of physical, emotional and moral silence, as in many of the pictures of Edward Hopper. Indeed, Hopper's work has drawn the epithet, 'metaphors of silence' (Renner, citing J. A. Ward in Renner, 85). Rolf Renner points out that 'just as all utterance is governed by what remains unsaid, and by silence, so too Hopper's art has its centre of gravity in what is not actually visible in the paintings…Hopper's pictures are about tension and isolation, and the silence indicated in many of his situations has major dramatic and communicative value in his aesthetic scheme' (ibid.). That in itself is an emotional dialogue. In short, the auditory response may be narrative, subliminal or conscious—or a combination of them all; we hear with our ears, but we listen with our minds.

Pictures at an Exhibition

Between October 2017 and March 2017, the *Fondation Louis Vuitton* staged an exhibition, *Being Modern: MoMA in Paris*, which distilled a collection of key works from New York's Museum of Modern Art, an institution established in 1929, enabling the preservation of much European art that would almost certainly have been lost in the turbulent years that followed. Works by Cezanne, Dali, Picasso, Kirschner and many others thus returned to the continent that bore them, and the 'voices' of the artists—individually and collectively—spoke and sang within the refreshing 'acoustic' of a changed environment. At a time of rising nationalism, the performance of this international choir was well timed. Indeed, the first catalogue in 1942 of the MoMA collection expressed such a concept explicitly in darker times: 'It is important when Hitler has made a lurid fetish of nationalism that no fewer than 24 nations other than our own should be represented in the museum collection' (quoted in Bajac, 22). In the Paris selection alone over one hundred artists from both sides of the Atlantic filled the galleries with the colours of sound and the sound of colour.

Visiting an art gallery or museum is an auditory experience even before specific works are examined; every room has its own acoustic, and the cumulative memory of such a visit is often retained as a sonic echo in the mind. The environment provides a series of circumstances through which the visitor walks, hearing the acoustic of place, ever changing as the crowds move and shift, and then focusing attention on individual art pieces in turn. The Russian composer Modest Mussorgsky, in his work, *Pictures at an Exhibition*, provides us with a clear musical interpretation of this experience, linked by a theme representing the *promenade* of the visitor between individual artworks, each of which Mussorgsky expresses in sound. Our overall sonic experience of a gallery provides the soundscape into which are placed the focused specifics of the art objects themselves, each with the potential to evoke their own sound worlds in the imagination; internalised sound placed within the context of physical sound, just as we 'heard' Samson slaying the Philistines earlier in the Victoria and Albert Museum, and as Keats 'heard' the grecian urn surrounded by the acoustic of the British Museum in which it stood. Leaving the room, we carry the memory of it: of the objects we have seen and the impression they made upon us, but also of the place itself. On the surface, it is a shared experience with other visitors, while in audio terms, every personal response remains unique and of our own making.

Franz Kafka said: 'Everyone carries a room about inside him. This fact can even be proved by means of the sense of hearing' (Kafka, 1). We may choose to purchase a catalogue as a souvenir of our experience; if so, when we open it in a new environment, say our home, office or classroom, the sound of this new place provides a changed audio backdrop, while the 'music' of the image may—or may not—remain the same as on first seeing/hearing. A broad analogy would be the experience of live music in the concert hall, contrasted with a recording of the same artist and/or work listened to within the environment of the home; one provides a direct witness to the event, while carrying with it all the unpredictability such a happening brings with it, while the alternative of a reasonably controlled situation, listening to the same work as a purely auditory experience, may aid contemplation, but through the medium of a copy. One could argue that neither is totally definitive, and indeed such a reading of the performer's original intention could only really occur were the performance itself to take place in the presence of the listener alone. Nevertheless, we find ourselves returning to experience the original in situ, in order to replenish our mnemonic sense of the event, and thereby nourish our relationship with the work itself. In other words, a painting performing 'live' is a happening of its own, but the sonic properties of the space itself as a part of experience are important in that event and its memory. A dry external acoustic compared to the liquid reverberations of a cathedral is a part of the performance of the world.

Were a composer to orchestrate a response to *Being Modern*, the sound forces would have indeed been various. The sheer range of work, historically and creatively on display, was as diverse as the decades represented, and every voice spoke of its time and across time. It was presented in three sections, from the earliest work of MoMA's first decade, through Minimalism and Pop Art, to the most recent acquisitions. There were indeed sound installations that spoke for themselves, such as Janet Cardiff's *Forty-Part Motet* from 2001, with forty loudspeakers, each a black box at head height, sounding Thomas Tallis's sixteenth-century 40-part motet, *Spem in Alium*. In this work, curator Quentin Bajac's description of the collection as 'polyphonic' (Bajac) was given literal expression, yet the word was applied to the exhibition as a whole and such a term was both appropriate and exact in referencing the imaginative and metaphorical murmur of sound generated not only by this exhibition, but by numerous others that may be experienced both externally and internally as a partnership between vision and sound in the mind. To take one example from *Being Modern* is to demonstrate the potential analogies between visual and sonic art, and our personal interpretation of them. Mark Rothko's 'No. 10' (oil on canvas), some seven feet by nearly five feet in size, painted in 1950, a series of horizontal blocks of colour, shades of off-white, grey, dull blue and yellow, is the epitome of Rothko's mature style, in which he left all vestiges of figuration behind in favour of solid fields of colour. In so doing, as Margaret Ewing has said, this work and others like them 'demonstrated the artist's transformation of his canvases into vessels for emotional response' (Bajac, 108). In 1943, Rothko and his artist friend Adolph Gottlieb sent a letter to *The New York Times* in response to a critic bewildered by the apparent lack of meaning in the artists' biomorphic abstract paintings: 'No possible set of notes can explain our paintings', they wrote. 'Their explanation must come out of a consummated experience between picture and onlooker' (Rothko and Gottlieb 1943, quoted in Bajac, 108). For the confused critics, one would hope that this was helpful advice; an even more useful guide might have been to encourage the viewer to allow the visual to touch the senses in the same spirit as music, bypassing the intellect and touching buttons of emotional response beyond rationalisation and interpretation.

The Hunters in the Snow

The concept of a piece of sound art has become increasingly accepted within gallery culture. In the summer of the year 2000, the musician and writer David Toop curated *Sonic Boom: The Art of Sound* at the Hayward Gallery, London. Toop was conscious of the context in which we hear sound: the inspiration for Akio Morita and Masaru Ibuka to develop the Sony Walkman had been, after all, the ability for listeners to avoid conflicts caused by competing sounds while listening to music. John Cage's response to listening had been the opposite. In his *4′33″* he had forced the audience to consider the sound of the room in which the so-called silent piece was 'performed'. Likewise, Toop, in his introduction to the Hayward Gallery exhibition, wrote:

> All the artists in *Sonic Boom* are alert and responsive to the richly clamorous environment in which we are now immersed. Rather than searching for

ways to cancel out the murmurings, hummings, pulses, whistles, alarms, signals, irritations, pleasures and shocks of the contemporary soundscape, they focus on their essence, impact and effect, so shaping new meanings for a bewildering range of aural events. (Toop, 15)

This is an acknowledgement of the sound environments within which we experience art, whether it be sonic, or apparently silent, just as, in John Berger's words, 'Photography is the process of rendering observation self-conscious' (Berger 1980, 19). And if it is so for photography, surely it may be said to be true of all visual art? What if maps could talk? What if the mapping of shape and place could be tethered to the cartography of thought and imagination? What if the earth's apparently silent voice could be translated into sound? In 2015, the AV Festival in the North East of England toured a sound installation by Susan Stenger, called *Sound Strata of Coastal Northumberland*. Stenger's 59-minute work was a sonic representation of a 12-metre long hand-drawn cross-section map of the coastal strata from the River Tyne to the River Tweed, created by a nineteenth-century mining engineer and cartographer called Nicholas Wood. Her work in this context is based on the sound of drones from Northumbrian pipes, a bed upon which other sounds—song, industry and imaginative abstract compositional techniques—riff and intertwine. In an essay accompanying the map, Wood referred to the area under his consideration, from Newcastle to Berwick-upon-Tweed in musical terms, as a 'suite of rocks'. Stenger in her turn gave terrain, geology and cultural history a range of voices that overlaid one another as do the strata of the earth's fabric. In other words, she 'read' Wood's 'score' imaginatively and articulated it in sound.

This is exactly what we do within our head when we read a poem or a book, a mental process that gives us the instrumentation to orchestrate the printed codes into imagery. In fact, the internal process goes further, turns three hundred and sixty-five degrees, because a visual or audio image itself as personal experience and circumstance to make a drama that in turn is mitigated by our own personality and placed in our memory bank. Gilles Deleuze has written: 'Musical art has two aspects, one which is something like a dance of molecules that reveal materiality, the other is the establishment of human relationships in their sound matter' (quoted in Stenger, 15). The miracle of composition is the revelation of patterns of sound placed on silence that touch a chord of recognition in us, as with our responses to pure, abstract visual tones. Wassily Kandinsky related colour to sound in many ways, akin to synaesthesia, speaking in one of his best-known writings, *On the Spiritual in Art*, he sought to define this idea:

> A sort of echo or resonance, as in the case of musical instruments, which without themselves being touched, vibrate in sympathy with another instrument being played…Our hearing of colours is so precise that it would perhaps be impossible to find anyone who would try to represent his impression of bright yellow by means of the bottom register of the piano… (Kandinsky, 158–9).

Kandinsky wrote his book in 1910, but even before the First World War, he was interested in the pioneering work possible in the potential partnership between audio and visual forms, in particular within the field of mental health: 'Various attempts to exploit this power of colour and apply it to different nervous disorders have…noted that red light has an enlivening and stimulating effect upon the heart…' (ibid.). He also noted that his contemporary, the composer Scriabin 'has constructed empirically a parallel table of equivalent tones in colour and music…Scriabin has made convincing use of his method in his *Prometheus*'.[1] (ibid.)

*   *   *

We are each of us composers, and our orchestra is our imagination. Stenger's sound work is rooted in a partnership with the visual, and we all possess the capacity in one degree or another to sonically incarnate and articulate any visual image presented to us. We can verbalise our response to it of course, describe it in language, or philosophise upon its content as did, for example, W. H. Auden on Brueghel's 'The Fall of Icarus' in his poem, 'Musee des Beaux Arts'. In such ways do we bring an object into the world and consciousness through new ways, giving a thing a life independent of its origins but at the same time dependent upon its original manifestation. Before this, however, the work has to 'speak' directly to our consciousness; to recognise its sonic potential is to develop layers of meaning that can greatly enhance its richness for us and the complexity of our response. Let us take as an example another painting by Pieter Brueghel the Elder, known variously as 'The Hunters in the Snow' and 'The Return of the Hunters'. Painted in 1565 as one of a series, five of which survive, depicting different times of the year. The original is housed in the Kunsthistorisches Museum, Vienna, and it is a useful example to discuss, partly because of its obvious sonic qualities, but also because it is one of the artist's most famous works, familiar to many and widely available through second-hand images.

The scene is set in the depths of a European winter, during December or January. In it, three hunters are returning with their dogs from what would appear to be an unsuccessful hunting trip. They trudge wearily through the snow, and their dogs hang their heads. One of the men carries the body of a dead fox, an indication of the paucity of their efforts, and there are the footprints in the snow of a small animal, possibly a rabbit, showing missed opportunities. They have come to the brow of a hill, and below them, there is their village stretching before them towards a strangely spectacular mountain range, clearly uncharacteristic of the rest of the view, which we may take to be perhaps Belgium or Holland. It is a still, overcast day, and the snow on the ground appears to be fresh-fallen. Skaters are moving across frozen ponds, playing hockey and curling; there is a frozen water wheel, birds swoop from the bare, leafless trees above the men's heads, and nearby several adults and a child are preparing food at a fire outside a wayside inn. It is all muted in terms of colour, and the smoke rises straight and hangs in the windless air. There is about it that particular stillness after snow in which every sound seems to accentuate by its silence, and distant voices ring out across the landscape. Absorbing the scene, we can identify with the cold that is implied visually, while allowing the mind to absorb the suggested sound within the painting, finding ourselves moving into a three-dimensional perspective that echoes the impact on the eyes. There is a soundscape playing within our heads that is parallel with the world depicted within the picture itself, muttered voices and whines from the hunters and their dogs in the foreground, voices and the crackling fire to our left from the inn, the cries of birds above our heads. Beyond that, most audible, is the sound that comes to us from below, the village and its skaters, borne up to us here where we stand with the hunters, through the still icy air. It captures that moment when, surmounting the brow of a hill, the sound of the scene that is suddenly revealed below opens up, and a kind of wide-screen stereo impression floods into the consciousness. It is a painting that rewards meditative study.

It is no coincidence that 'The Hunters in the Snow' has featured in a number of motion pictures. Lars von Trier's film, *Melancholia* (2011), contains its image, as does Alain Tanner's *Dans la ville blanche* (1983) and it was the inspiration for Roy Andersson's 2014 film, *A Pigeon Sat on a Branch Reflecting on Existence*. Most significant of all, because of the poetic use of sound in all his work, is its presence in *The Mirror*, and *Solaris* by the great Russian director Andrei Tarkovsky. Notably in *Solaris*, in which several of Brueghel's seasonal paintings are depicted on the walls of the space station, the sonic aspect of 'The Hunters in the Snow' contributes to the presence of nostalgia for the absence of earthly humanity. The camera lingers over details in the picture, and we hear the whispered sound of its story: the trees, birds, dogs and footfalls on snow. The landscape of earth is brought into the sterile environment of the space station, and it symbolises a longing for another place, just as Brueghel himself may have been suggesting, at a time in the 1560s of religious revolution in the Netherlands, a reference to an ideal, or an idealised past time in rural life. This year, 'The Hunters in the Snow', will once again appear on Christmas cards around the world, and—as in many of Brueghel's paintings—its suggested sound world will speak, perhaps subliminally, in parallel with its visual message. The exercise of exploring a painting in terms of its sound can be a useful one and may be applied to an infinite range of images; I would suggest that anyone interested in pursuing this line of enquiry selects their own examples, thus demonstrating through personal experience the imaginative music held to a greater or a lesser degree within otherwise seemingly mute visual objects, when the mind directly engages with them.

Studium and Punctum

In the summer of 2015, The National Gallery, London staged *Hear the Painting, See the Sound,* an exhibition in which six noted musicians and sound artists generated soundscapes to accompany a painting of their choice, drawn from within the Gallery's collections. The word 'staged' is appropriate in this context; each of the paintings selected were bathed in a spotlight, surrounded by subdued twilight, and the aural accompaniment enabled the viewer to linger in front of each of the works, experiencing them almost as theatre, while the artists and composers chosen for the project each had distinguished pedigrees in their respective fields. Natural history sound recordist Chris Watson selected Akseli Gallen-Kallela's 'Lake Keitele', Susan Philpsz chose Holbein's haunting and mysterious picture, 'The Ambassadors', and Janet Cardiff and George Bures Miller interpreted 'St Jerome in his Study' by Antonello da Messina. The American composer Nico Muhly used the fourteenth century 'Wilton Diptych' as his subject, Jamie xx of the electro duo, The xx selected Théo Van Rysselberghe's 'Coastal Scene', and the French film composer Gabriel Yared created a score to complement 'Bathers' by Cezanne. In introducing the project, National Gallery director Nicholas Penny stated that 'when sounds have been composed in response to a work of art, they can encourage – even compel – concentration'.[2] While this may be true, the concept of an imposed sound commentary to a work of art is problematic, in that it can intrude and negate the viewer's personal sonic interpretation, as well as potentially the intention of the artist themselves. Reviewing the exhibition for the *Daily Telegraph*, Mark Hudson's view was that 'a painting should generate its own music, its own soundtrack in your head that is entirely personal to you. This experiment is an engaging novelty, but it essentially limits the viewer's response'.[3]

The issue here is directly linked to our personal interpretation of works of art, and in particular to specific details that may draw the eye (and imaginatively, the ear) and which may vary with each individual.

In his book *Camera Lucida*, Roland Barthes explored this idea in relation to photography, turning the eye of the beholder back upon him or herself as an exposition of their own mind: 'The photograph is literally an emanation of the referent. From a real body, which was there, proceed radiations which ultimately touch me, who am here' (Barthes 2000, 80). In moving from painting to photography, we take a highly significant step, away from considered interpretation, to a direct and immediate interplay between the human observer and the moment itself. In so doing, we come closer still to imaginative sonic response. Barthes, in developing his ideas of the mind's interplay with photographic images, employs the terms 'Studium' and 'Punctum' as concepts of personal response. In order to utilise these words ourselves before proceeding, we must identify them within the context of a picture. The photographer, George Powell, sees Barthes view of Studium as 'the element that creates interest in a photographic image. It shows the intention of the photographer, but we experience this intention in reverse as spectators. The photographer thinks of the idea (or intention) then presents it photographically; the spectator then has to act in the opposite way – they see the photograph, then have to interpret it to see the ideas and intentions behind it'.[4] Thus the studium belongs to the photographer and grows initially from the intention behind the photograph. (One could significantly also apply this to the work of the sound recordist during the development, for example, of a radio feature or similar.) The punctum, on the other hand, as Powell suggests:

> …is an object or image that jumps out at the viewer within a photograph. Punctum can exist alongside studium, but disturbs it. Punctum is the rare detail that attracts you to an image. Clearly this second element is much more powerful and compelling to the spectator, changing the 'like' of studium to the love of an image. As a photographer, an understanding of punctum could potentially allow me to make stronger images, although I feel that punctum needs that accidental quality about it to be most effective, because it is so personal and could be different for everyone. Basically it could be anything, something that reminds you of your childhood, a sense of *deja vu*, an object of sentimental value. The punctum is very personal and often different for everyone.[5]

Thus the key element in our relationship with an image—visual or aural—is the response that occurs within us, and the mistake we may make in our assessment of how that relationship operates, lies in a failure to appreciate that seeing and listening are not passive, but active and creative acts. As Gaston Bachelard says, 'contemplation is essentially a creative power' (Bachelard, 49).

It also becomes clear that the punctum in a picture, the detail that draws our attention and may obsess us to the point of defining the whole picture, can be auditory as well as visual. Just as Roland Barthes found himself focusing in spite of himself on a part of a picture—say the belt on a woman's dress or a pair of shoes—that same defining detail might also evoke the idea of a sound in the viewer's mind. To return to the example of Brueghel's painting, there might be an auditory punctum evoked in a visual punctum: the fire to the left-hand side of the action in 'The Hunters in the Snow', for example, may be a 'heard' image as well as a seen one. Likewise the birds flying over the hunters' heads are surely calling to one another? This idea in turn leads to another, because sound is temporal: both a visual image and a recording play with time and space, although in different ways, in that as we have already discussed, a photograph freezes a moment in time, while the sound we 'hear' as we look at it moves through time, walks alongside our consciousness for a while. A photograph differs from a painting, in that it is captured in an instant, whereas a painting is an act of deliberation and prolonged observation of an object or scene that is changing as it is observed, and yet which ultimately becomes frozen in the finished work of art. When the work of the observer begins, both photograph and painting are subjected to objective interpretation, and the evocation of sound is a part of that observation. We interpret the world sonically through a progression of moments, and contemplation takes many forms; at root is recognition of a situation, partnered by memory that touches a poignant chord of need in the human spirit. Remembering is crucial to consciousness, integrating the past with the present, as John Locke wrote: 'Consciousness, as far as it can be extended…unites existences and actions very remote in time into the same person' (Locke 213). It is in the partnership between the photograph, an image that is both metaphor, and medium for memory, and the passing invisible messages provided by sound as both commentator and image-maker in its own form, that memory and consciousness may often find a most profound internal response.

Light. Space. Time.

Thus the relationship between the apparently autonomous visual image and the sound that it evokes through suggestion in the mind is more complex than we might at first imagine. We have long understood that a sound can evoke a visual reverberation of its meaning, so it should come as no surprise to realise that the reverse is also true. It is the concept of time as an element in the making of both image and sonic signal, however, that we must consider if we are to gain a fuller understanding of relationships and differences. To study 'The Hunters in the Snow' is to absorb a fixed image, but the sounds of the moment captured affect the imagination in real time. A photograph is an even more precise, fixed record of a moment, in the sense that the device that recorded it responds to the will of the photographer in an instant. A camera, we should never forget, is after all, a machine, just as is a digital sound recorder. The *idea* of a place—its sonic presence manifested as bird song, trees stirred by the breeze, the crunch of an approaching footfall on the gravel path, sounds from the house beyond—all of these can offer a soundtrack to the silent instant of the photograph. It offers a kind of poetic truth, belonging exclusively to the individual viewer of the picture, even independently of the visual image itself.

Sound is a metaphor for our mortality, because it is always disappearing. The loudest sound only emphasises the silence that surrounds it, so in its bleakest incarnation, sound is a *memento mori*. Taken this way, the tolling of a single bell, as expressed at the start of this book, that may be seen on one level as a bridge between the material world and that of the spirit and/or the imagination, may also be heard as analogous for the sense of life vanishing gradually into death. To examine a painting related to social or industrial history, for example, one of the North of England industrial cityscapes by L. S. Lowry is to hear the bustle of the crowd on their way to work, the hoot of the factories and the murmur of the town beyond, while at one and the same moment being reminded—if we choose to remember—that these places have now changed, these times are gone and these people—both those represented in the image, and indeed the artist himself—are all dead. These things are poignant precisely because they can unexpectedly reach out and touch our consciousness 'like the delayed rays of a star' (Barthes, 81).

We should continue to remember how relatively recent is the electronic technology of the media that allows the clues of memory to be held. The strangeness of the invisible medium of sound, the almost supernatural invisibility of wireless signals and voices through the air, was to a degree 'earthed' by the development of recording, fixing the memory of sound and pulling it back from the darkness of death, and contributing to the scientific revolution that helped to make human memory more self-conscious, reminding itself of its presence in the shaping of lives. Both a photograph and a sound recording have the capacity to move us from the moment as it is lived, into memory. Sound itself possesses the facility of imaginative photography, and the most eloquent (in every sense) example of this may be found it our capacity to absorb music. The explanation of this is linked to the mystery of music itself and its relationship to time, which is constantly moving through a permanent present. Sound—and in this case music—is effectively time made audible, but the experience of it is constantly current. The brain absorbs a newly heard piece of music on both a conscious and a subconscious level, and when this occurs, we are very actively absorbed in learning, processing information that we then store, pending retrieval. As Oliver Sacks has said: 'When we "remember" a melody, it plays in our mind; it becomes newly alive…We recall one note at a time and each note entirely fills our consciousness, yet simultaneously it relates to the whole' (Sacks, 227).

As we listen to music, we are hearing the moment, while relating the moment to the immediate previous moment, and moving within a split second into the next moment. It is hearing, having heard and being about to hear in one and the same instant. It is in other words, very close to the experience of having our attention caught by striking photographic image. We might take this further into the realm of language and consider the images that words evoke pictorially through speech, including meaning but also through timbre, pitch, pace and volume, and in this context view the printed word as a kind of musical notation, a record of thought that manifests itself through the eyes, being absorbed by the brain and then translated imaginatively into sound, either spoken aloud or imagined internally. It is the partnership between sound and image—the sonic responses produced in us by art and conversely the pictures that sound makes—that places us in the world. Once that response has been set up, it has the further capacity to become lodged in memory, thus evoking an image of itself and often, of one's first experience of it. We are ourselves part of a huge audio–visual work, and while Berger may say 'It is seeing which establishes our place in the surrounding world' (Berger 1972, 7). We might add, 'and listening'.

Just as a sound can transport the imagination across the globe, or even in imaginative terms, the universe, a musical phrase or a harmony has the potential to move the mind through a lifetime, or even through centuries. Reflecting on the many-voiced chorus that sang through the rooms of *Fondation Louis Vuitton* in Paris during the MoMA exhibition during the winter of 2017–2018, the visitor is drawn back to a memory of the last work in the exhibition, Janet Cardiff's 2001 sound installation, *Forty-Part Motet, Spem in Alium*. It is hauntingly familiar music, redolent of the spaces for which it was composed, and the first impression in a gallery context is of the technology, the forty speakers on stands, in a circle, singing to one another across a bare white room. Beyond that, however, there is the sound of late sixteenth-century churches, the great spaces in which this music once echoed, and in which even today it finds its truest expression. It is music created in partnership with sacred architecture, belonging to a time and world visited in Chapter 3 of this book, and when heard, the image must surely be of the towering pillars of the great cathedrals of Europe. The interweaving of the voices is seamless; the visual art celebrates Tallis's genius, the polyphony of his day, but also the buildings themselves, and the art within the shape of them. Cardiff's work, through sound, evokes the memory of architecture, much as the reconstruction of performance in Istanbul's Hagia Sophia, discussed earlier provides an imaginative sense of the physical space itself. At the same time, as it sends own message from its inward-facing speakers, seeming as they do to discourse together, it is the metaphor that rings most strongly, that of voices in harmony, transcending a gallery space and transforming it into cathedral arches, vaults and vistas. Paul Claudel, recalling a visit to Strasbourg Cathedral, remembered that 'the great empty spaces, whose relationship one with another constitutes the religious matrix and the vessel of our common presence in God, have as their role only the modelling of the invisible about us, the circumscribing of the volume of air and of thought suitable for our nourishment, this breath that God places at our disposal to draw on, to transform into word and song'. Whether or not we consider such experiences to be religious, it is hard to deny that sound and art (and sound *as* art) have as their common denominator, aspirations to express our desire to preserve time and hold the passing moment, be it in a painting, a sculpture or a building. It may be transcendent, but it is also intensely and profoundly human in origin.

The artistic image may evoke peace or discord, tranquillity or violent conflict; we may visit it in a gallery, or it may come to meet us, taking us by surprise on a seashore or hilltop; wherever we interact with the man-made expressions of ideas, the presence of sound either as the artwork itself, or as an idea suggested by it literally or tangentially in the mind, enhances the experience of the visually creative world and perhaps changes it while adding layers of meaning. As Picasso's *The Old Guitarist* (1903) challenged the art world by flattening and fragmenting pictorial space, and as when in the Second Viennese school, composers such as Schoenberg led music into new realms of sound, so might we bring our personally created soundscapes to what we see, finding new and strange sonic worlds within worlds, and like the poet Wallace Stevens and his 'Man with the Blue Guitar' come to understand that things may not quite as they are…or seem to be.

Notes

1. *Prometheus: The Poem of Fire* (op.60, 1910) by Alexander Scriabin was first performed in the same year Kandinsky wrote his treatise. Part of its original orchestration was the employment of a 'clavier à lumières' or 'Chromola' (a colour organ, invented by Preston Millar.) Today, this instrument is rarely featured in performances of the piece.
2. *Daily Telegraph*, 7 July 2015.
3. Ibid.
4. George Powell, blog, 2008. https://georgepowell.wordpress.com/2008/07/01/studium-and-punctum/.
5. Ibid.

9

Writing Aloud: The Sound of Voice on Page and Social Screen

In March 2009, Donald Trump activated his Twitter account, @realDonaldTrump. By July 2018, he had gained 53.2 million followers and tweeted more than 38,000 times, and all the major world news outlets routinely embedded his messages within their online news reports. In a way not seen before or elsewhere, the president of the world's most powerful country utilised social media as a vehicle for his opinions and political announcements. The medium allowed him to make unilateral statements at any time of day or night, and in whatever state of mind or body pertained at a particular moment, and to by-pass advisers and the policy-making machine of Senate, Congress and even staff in the White House. He would frequently make comments on controversial issues such as the US travel ban, transgender military recruits and immigration. Donald Trump was only one of thousands of political and public figures using Twitter accounts, but the scale of his activity—on average an output of ten tweets per day in mid-2018— made the volume of his usage virtually unique. It was not only the sheer output sent from a mobile telephone, night and day, that is interesting, but the *tone* of *voice* 'audible' within its content. It was unmistakably the voice of Donald Trump, and his audience recognised it through its ubiquity on broadcast, cable and internet media globally and

© The Author(s) 2019
S. Street, *The Sound inside the Silence*, Palgrave Studies in Sound,
https://doi.org/10.1007/978-981-13-8449-3_9

instantly. The Twitter feed of the American president was just one product of the Trump brand and was identified as such in the same way as one recognises a familiar popular song or anthem. It may also provide us with an example of how, with social media, we are returning in some ways to a spontaneous representation of the heard voice through text.

Robert Frost's theories of 'the sound of sense', his expression of what he called 'the abstract vitality of our speech' (Frost, 80) is highly germane to the consideration of the communication of voice, mood, temperament and feeling in the written word, whether it be on the page, or on the screen, in particular with respect to social media, where the message is written, sent and received in almost an instant. A technical manual is made up of language without character, but our personal communications often convey a sense of who we are as much as what we are actually saying. In Frost's words, 'there are tones of voice that mean more than words. Sentences may be shaped as definitely to indicate these tones' (ibid., 204). This philosophy of speech sounds transferred to the page informed Robert Frost's writing from an early age and was to influence many poets who came after him. In March 1915, he wrote to a correspondent:

> My conscious interest in people was at first no more than an almost technical interest in their speech – in what I call their sentence sounds – the sound of sense. Whatever these sounds are or aren't (they are certainly not of the vowels and consonants of words nor even the words themselves but something the words are chiefly a kind of notation for indicating and fastening to the printed page) whatever they are, I say, I began to hang on them very young. (ibid., 158–9)

If the key to the sound of a thought transferred from the mind to a form of written expression is not in the structure of the words, it would seem that tone is the quality that communicates who we are, ourselves 'spoken' as it were directly as we write. As always, it comes down to listening; meaning is made in the process of interpretation as well as at the point of utterance. We are free to understand or misunderstand according not only to what is said, but in response to the tone in which the statement is expressed, or how we *consider* it to have been expressed. A Twitter message can be deleted, but its inference cannot be unheard, and whereas a conversation spoken face-to-face can move through phases governed by caveats, corrections and rephrasing of arguments, electronic correspondence may cause offence or lasting damage if it is misread or misunderstood at first sight. On the other hand, it can be used to devastating effect when the tone is deliberately employed to raise emotions, for example by the SHOUTING OF CAPITAL LETTERS within the text. Every social media innovation teaches us new rules and conventions. Frost used italics sometimes to make a point, as in a letter to his friend, John Bartlett in February 1914: '*The ear does it*', he wrote, ensuring the page conveyed the emphasis that signalled the importance of the words within his own mind; 'The ear is the only true writer and the only true reader…Remember that the sentence sound often says more than the words. It may even as in irony convey a meaning opposite to the words' (ibid., 113).

At approximately the same time as Frost was developing his ideas on the sound of sense, the Russian painter Wassily Kandinsky was coming to not dissimilar conclusions of his own relating to word sounds. The power of inner meaning, the deep root from which language grows, fascinated him and was expressed in a sequence of poems and accompanying prints which he called *Sounds*. Elsewhere in his writing, he explored the subject specifically:

> Words are inner sounds. This inner sound arises partly – perhaps principally – from the object for which the word serves as a name. But when the object itself is not seen, but only its name is heard, an abstract conception arises in the mind of the listener, a dematerialised object that at once conjures up a vibration in the "heart."…Skilful use of a word (according to poetic feeling) – an internally necessary repetition of the same word twice, three times, many times – can lead not only to the growth of the inner sound, but also bring to light still other, unrealised spiritual qualities in the word. Eventually, manifold repetition of a word…makes it lose its external sense as a name. In this way, even the sense of a word as an abstract indication of the object is forgotten, and only the pure sound of the word remains. (Kandinsky, 147)

In a negative sense, we may grow overused to words in frequent media usage, to the extent that their significance is dulled, and our response to them blunted. At the same time, we all have our own turns of phrase that may act for others as a shorthand in terms of recognition on the page, computer or telephone. We each of us present a range of 'selves' to the world. At the root of our presentation of self may lie our true being, but the image can be very different, and this is conveyed in a range of voices, both physical and virtual. Speech is complex and subtle, a learnt skill to transmit ideas and opinion. Michael Argyle, in his book, *The Psychology of Interpersonal Behaviour* wrote: 'Linguists sometimes present language as printed words on paper. This is a mistake: the real unit is the utterance by one individual to one or more others, in a situation, in a conversational sequence, where he or she is trying to influence the other' (Argyle, 56). Argyle was writing in 1967, long before the social media revolution that created an interconnected world, and the capacity for speech and voice to be conveyed virtually. As so often with technology, people—particularly the young—possess the ability to subvert it with the development of new informal means of social expression that enhance the message, add emphasis and employ new ways of demonstrating a point of view. Thus, the abbreviation of emotions into acronyms, emoticons and the use of lower and upper case to underline emphases means that we do indeed come closer to the simulation of an opinionated voice. What is lacking within the form is the grain, timbre and tone that might mitigate meaning. The expression for instance of laughter—a spontaneous act—through letters, words and emoticons, implies a premeditated act that is at odds with the reality and gives little sense of the depth with which the emotion or response is actually felt. As usage has developed, strategies have evolved through which the artificiality of the medium has been adjusted to the speaking voice, simulated through a vocabulary of symbols and typographical tools.

Facebook, Instagram, Twitter and other social media platforms are full of 'friended' or 'unfriended' people who have never met, while lifelong relationships have been made when two voices touch in empathy. The voice of criticism may be benign, constructive or aggressive, and it has always been so; Samuel Johnson and Dorothy Parker possessed voices that transmitted eloquently through their writing, and the tone of their thought owns a character which may or may not reflect the actual sound of that person when engaged in speech. In some cases, a writer may be eloquent when at their desk, and yet tongue-tied in company. There is nothing new in this; the 'shouting voice' of opinion has manifested itself through formal critical history. Within today's world of social media, opinionated, prejudiced and/or ill-informed voices may present a front to a virtual society to which they contribute, which may well be very much at odds with the real personality of the writer. At the same time, legitimate statements and educated views may battle to be heard or become themselves interpreted as 'fake news'. Language—in all the available forms of expression—becomes a weapon of bullying, negation and intellectual attack. As Ruth Page has written in the Palgrave book, *The Language of Social Media*, 'the ambiguity associated with online representation of the self sits within a wider complex of debates about the nature of authenticity, trust and reputation that are crucial to the discussion of impersonation in online contexts' (Page in Sergeant and Tagg, 47).

Hoaxes and impersonations are by no means limited to online contexts. Examples date back centuries, including literary and non-literary hoaxes. Of the former, one of the most notorious may be seen to be the young Thomas Chatterton's brilliant forging of the Rowley Poems, at the age of sixteen simulating not only the antiquity of the medium, but the very voice of another time. (We have also already considered another example of this in the form of James MacPherson's 'discovery' of the supposedly ancient text of *Fingal*.) Today, Facebook may provide a platform for both violent debate and light-hearted exchange; research shows friends adopting comedy personas with imagined voices to create dialogue that may ultimately transfer itself to other media in the form of formal text or entertainment. The virtuality of voices that transmit from mind to mind and their effect beyond the physicality of speech enables global conversations between people who often may never meet. If or when they do so, sometimes reality can reveal very different personal relationships.

The Written Voice

During the late summer of 1820, John Keats was mortally ill, with tuberculosis; it had been decided that he must be moved from his home in Hampstead, London, to warmer climes. Accordingly on 13 September, he and his friend Joseph Severn left for Italy, where, in a villa on Rome's

Spanish Steps, he was to die on 21 February 1821. We may see these last, tragic months through a gauze screen of sentiment, coloured by the romantic poetry with which he gifted literature, and the melancholy of the death of a young (24 years old) man, in the prime of his life. The literary imagination does its work, and we create our picture of the dying Keats in terms of languid physical beauty and a soft, gentle, perhaps 'standard English' voice, submissive and with the docility of chronic infirmity. Thus, it might come as a surprise to us to read a postscript to a short letter he wrote on 14 August 1821—less than a month before his departure for Rome—to his friend and publisher, John Taylor. Keats is busy arranging things for the journey, including the choice of shipping route. He had received a letter from his friend Percy Bysshe Shelley, already in Italy, on 27 July, who had mentioned the possibility of coming via Leghorn, and he is here asking Taylor to investigate this. It is clearly a hurried note:

> I do not think I mentioned anything of a passage to Leghorn by sea? Will you join that to your enquiries, and if you can, give a peep at the birth, if the vessel is in our river? (Rollins ll, 318)

The brief message has all the vernacular and hurry of someone in the throes of making final travel arrangements; these days it would probably have been contained within a telephone text. Then comes a P.S., which is longer than the note itself. Keats is packing, and as is always the case, overlooked things are coming to light:

> Somehow a copy of Chapman's Homer, lent to me by Haydon, has disappeared from my lodgings – and it has quite flown I'm afraid, and Haydon urges the return of it so I must get one at Longman's and send it to Lisson Grove – or you must – or as I have given you a job on the river – ask Mistessey. I had written a note to this effect to Hessey some time since but crumpled it up in hopes that the book might come to light. This morning Haydon has sent another messenger. The copy was in good condition…- Damn all thieves! Tell Woodhouse I have not lost his Blackwood. (ibid., 318–9)

There is something endearingly colloquial about the voice in this panicky, irritated letter that is instantly recognisable as a situation in which any of us might find ourselves today. How often, while tidying, or clearing out effects, have we come to the realisation that something loaned to us cannot be accounted for? Added here is the urgency prompted by the fact that his friend, the painter Benjamin Haydon, knowing Keats was shortly to leave London, almost certainly for the last time, wants to get his book back! Here, we have the everyday voice of a writer who we usually 'hear' through the medium of his published poems. It gives us the man, with all his annoyances, frustrations and urgency as he prepares to go on his journey. John Keats was born in Moorgate, London. By all accounts, he had a London accent of the time, and he was a practical man. His early medical training had made him a realist about his own health problems, and he had a quick temper. His friend, Benjamin Bailey, mentioned however that when reading poetry, he adopted another voice, having 'listened to him "recite, or chant" Chatterton "in his peculiar manner"' (Bailey, quoted in Motion, 192). It is not unusual for some poets even today to affect a 'poetry reading' voice at odds to their own conversational tone, so it is interesting to note that this practice also seems to have been the case during the first half of the nineteenth century. The very fact that Bailey refers to this 'chant' as being 'peculiar' suggests very clearly that Keats's own voice was of a different sound and colour. What throws the expression of Keats's frustration in the letter into greater relief is a comparison with one of his most famous sonnets, which has as its subject, a copy of the book under discussion in the letter, although 'On First Looking Into Chapman's Homer',[1] written in 1816, when he was just 19-year-old:

> Much have I travell'd in the realms of gold,
> And many goodly states and kingdoms seen;
> Round many western islands have I been
> Which bards in fealty to Apollo hold.
> Oft of one wide expanse had I been told
> That deep-brow'd Homer ruled as his demesne;
> Yet did I never breathe its pure serene
> Till I heard Chapman speak out loud and bold:
> Then felt I like some watcher of the skies
> When a new planet swims into his ken;
> Or like stout Cortez when with eagle eyes
> He star'd at the Pacific—and all his men

> Look'd at each other with a wild surmise—
> Silent, upon a peak in Darien. (Keats, 38)

The point of the comparison between letter and poem is to demonstrate an ability to adopt an appropriate 'voice' for a particular medium. Today's poets may strive to make the sound of their poetry more akin to conversation; in writers such as Kate Tempest or Linton Kwesi Johnson, to name but two among many, we may hear the poet themselves speaking, representing language within the context of their time and culture, either personally, or through a character. Both the physical voice used for reading the words and the words themselves have changed; Keats may have adopted a 'poetry voice' to read, but likewise, the accepted 'voice' of the text itself was that which observed the prosodic rules and norms of the literary world of the time, and the 'voices' he was reading from his contemporaries provided a context for the sound of his own poetry. It was not until the early twentieth century that poets such as Robert Frost made inroads into a poetic language that by that time had become archaic to the ear, when the 'adopted voice' had come to be outmoded and inappropriate. As we discussed in Chapter 7, the written word is not the root of language, but a relatively modern visual expression of it. To quote Walter Ong again in this context:

> We are inclined to think of words as records because we are inclined to think of them as, at their optimum, written out or printed. Once we can get over our chirographic-typographic squint here, we can see that the word in its original habitat of sound, which is still its native habitat, is not a record at all. The word is something that happens, an event in the world of sound through which the mind is enabled to relate actuality to itself. (Ong, 22)

This becomes highly relevant when we consider the ever-developing field of social media, and our utilisation of new shorthand that uses a wide range of symbols, animation, illustrations and typographical subversions in order to transcend the written word in the expression of ideas, moods and affinities within limitations of textual space and time, often approximating as near as possible to the sound of ourselves at a given moment.

The Twittering Classes

The dramatist or novelist has the ability to create characters that speak with a reality that belongs to them, while having been invented by the writer. This is a projection of the sonic imagination. Within social media on the other hand, we have the capacity to adopt a voice other than our own, to as it were, reverse the process of Keats's writing and to make ourselves sound universal to the time, rather than particular to our immediate being. In short, we may project dramatisations of ourselves to the world, constructions of an ideal persona with which we can pursue dialogues with strangers. Shyness or an introvert or damaged personality can hide behind a self-created mask of language generating a 'voice' that is imagined by those with whom we interact on this virtual basis. Keats was writing to one other person; he would never have imagined that the words in his private letter to Taylor would ever be read by anyone else. As Chambers points out:

> This digitised era is the first in which personal connections of friendship become formalised through online public display. The question is whether this emergent ritual of displaying non-familial as well as familial social connections online affects conventional meanings and values associated with "friendship" and "intimacy." Questions about the intensity and speed of self-disclosure online, the unforeseen side effect of constant self-disclosure and how to sustain digital connections, are issues that provoke questions about the sorts of skills now required to be "a friend". (Chambers, 5)

Today, we may pour out our hearts to thousands of barely known unmet 'friends', place a family photo on Instagram or fire off sometimes random thoughts into the ether via Twitter. We do this as we engage with the world, minute by minute, part of the rush and hurry of life; it is instant communication and response. Thus, it is hardly surprising that the shorthand to depict our voice and personality has become more and more a part of the medium in which it is used. Chambers reminds us:

> The rise of social media has coincided with the introduction of several new words in the English language such as to "friend" or to "unfriend" a person; "offline friends" and "non-friend." The term, "frenemies" is used in the

context of online stalking: "stalking your frenemies." The term "unfriend" was selected as the Oxford Word of the Year in 2009…. (ibid., 6)

Use of social media is highly generational, and the voices reflect this. Teens have long sought to disguise meanings from adults, and the 'hip-talk' of the 1950s was one means of creating a kind of elite linguistic 'club' that excluded all who did not belong to it and understand the rules of its vocabulary. Likewise it was youth that led the way with the adoption of social media, and in particular its shorthand. As is usually the case, older generations followed the example of the young, and thus more extreme codes became necessary. Perverting the intended use of the keyboard, today old and young alike use symbols such as punctuation marks to depict 'smiley' or angry faces, bewilderment, fear, laughter and tears. Somewhere in this process, the use of capital letters became synonymous with the shouting voice and the exclamation mark as an expression of the raised voice of frustration or disagreement. Moving beyond language itself, the development of the emoji and the gif offers the opportunity to express these and other emotions and expressions of personality in various entertaining ways. We need these devices partly because they are shorthand, and partly because on social media, it is easy to be misunderstood. This is particularly true of humour, because it is a very personal part of who we are; sometimes the tone of voice is all. Social media may on the face of it, be about 'friends', but it is notoriously easy to give offence, particularly when one is addressing one's words to a largely global audience of thousands. The virtual voice in social text, however it is suggested, is physically unheard and therein may lie the issue. Seeking to write a conversation, we forget that in a real debate, the tone, pitch, volume and timbre are heard and contribute to the meaning. This is of course true of all text; the letter, the book or the poem, on the other hand may be much more considered and subject to revision prior to committing them to the gaze of others. We know that letter writing has declined, supplanted by emails and texts to a large degree, yet we continue to make or replicate sounds, however, the means may change. 'The basic elements of language are physical', wrote Ursula Le Guin: 'the sounds and silences that make the rhythms making their relationships' (LeGuin, 1). Replicating that or reading the virtual voice may be fraught with misunderstandings.

On the other hand, Twitter or Facebook correspondents who have an offline friendship in parallel with their online communications 'hear' the tone of voice that confirms identity in their close associates. Risks inherent in many areas of internet use involve understanding when a message is genuine or otherwise: is this really from who it purports to be? With familiar voices, we have almost instant recognition, and so it may be with a fake Twitter message. We might say, 'I thought something was wrong because it just didn't *sound* like you'. When Ruth Page, a Reader in the School of English, had her Twitter account compromised, a fake message, purporting to be from her, was sent to her followers. For a number of these genuine correspondents, it was the tone of voice and the use of grammar that betrayed the suspect nature of the communication, many interpreting…

> …a dissonance between the grammatical style that they usually observed in my updates, or judged appropriate for me based on my position as an academic…others said, "I'm familiar enough with your presence on Twitter that I trust you would be capable of constructing a properly worded message – the poor grammar was an immediate red flag." Other recipients detected a style 'that was not in keeping with British English,'" hearing in the text what they suspected might be an American accent. 'If a text appears to be stylistically at odds with expectations generated from contextual knowledge, then inauthenticity may be suspected'. (Page in Sergeant and Tagg, 56–7)

The sound analogy is clear: when we are familiar with a voice, either in audio or textual terms, we become highly tuned to variants in tone, and the subtleties with which we express ourselves become evident. We know one another better than we think we do, and as we write, we place our voice on the page or screen, creating ourselves word by word, either as who we are, or as who we wish to be perceived to be. Yet even when we seek to generate an idealised persona for ourselves, there remains a tone of voice that is not easy to impersonate. Within these parameters, we may adopt an appropriate voice, depending on the person or group with which we are conversing, much as we might in our day-to-day dealings with work colleagues, officials, or casual conversational situations in bars or restaurants, etc. 'The implication of an invisible, semi-public, and converged audience is that Facebook users are constructing their audience as much as they are targeting it. By varying the audience design strategies they adopt, users not only respond differently to the audience before them but actually put together or construct an audience from amongst the wider potential readership…' (Tagg and Sergeant in ibid., 169).

It is notable here that what many see as a network of friends has developed into a relationship between performer or orator and audience, and this reflects how social media users vary in their imaginative use of the medium. While some see it as pure interactive conversation—the virtual equivalent of an informal discussion around the kitchen table, others may use it as a platform from which to deliver 'truths' about themselves or the world. The alert reader can 'hear' the shift in tone and projection.

The instant nature of social media offers itself to hasty and sometimes ill-considered responses in interchanges; it is here that the raised voice becomes most evident, and sometimes, the uninvolved reader may feel themselves to be witness to an encounter that, had it been physical rather than virtual, might well have resulted in violence. In the era of letter writing, the author of a contentious communication might well have been exhorted to 'sleep on it' before sending. Allowing time for reflection, a calming of emotion, and revision is less likely in the heat of an exchange that is as fast as if the words were blurted out face-to-face. Most poets who are serious about their craft read drafts of new work out loud, then leave it for later reflection; infelicities of tone, unintended meanings and inadequate words or phrases in this way very often declare themselves eloquently.

On the other hand, for many, the very character-limit to the Twitter medium can prove to be a creative stimulus; we may be thrown back on the power of suggestion, impressionism, less may be more, and concision can be everything. Originally, authors were limited to 140 characters for each tweet, which was later increased to 280 characters. That said, the most common length of tweets is much less, generally something just over 30 characters, although the doubling of the character count has had an impact on how people write—or 'speak'. With its compression and discipline, Twitter has the potential for being the most poetic and eloquent of all social media forms, almost a modern equivalent of the haiku, with the capacity for depicting the true voice of its author. This may be the largest reading audience a person can ever expect to attain, a democratisation that at once permits direct and instant expression, and through it the conscious projection of a self. Such a 'self' might be either real or constructed, but either way, it has the potential to provide opportunities for a form of self-advertisement cloaked in the apparent conversation between friends within a specific group.

A letter is read by the recipient, as a voice speaking on the telephone is heard. A traditional radio broadcast, that is to say transmitted from a location and heard on a receiver, is tuned into, either by accident or design. Likewise, the operation of social media requires to be considered within a similar context. In order for a voice to communicate, it must be 'heard'. In 2009, Jay Rosen, professor of journalism at New York University, conducted an informal poll amongst his 12,000 Twitter followers, the question being, why they used the service? Responding to them during analysis of the experiment, he wrote that a 'surprise finding from my project is how often I wound up with radio as a comparison' (Rosen, quoted by Crawford in Sterne, 82). Some might say, a decade or so on from this project, that today we could suggest podcasting as a more appropriate analogy, and yet the older means of accessing, by 'tuning in' is actually more appropriate in this context. Rosen cites a Microsoft Network editor, Jane Douglas, who sees 'Twitter like a ham radio for tuning into the world' (ibid.). Kate Crawford's development of this idea in her article, 'Following You: Disciplines of Listening in Social Media', is illuminating:

> The act of "tuning in" reflects part of the process of engagement with social media. A Twitter user follows a range of people, some of whom will post updates that offer useful advice, amusing anecdotes, or interesting links. But many messages will simply be scanned quickly, not focused on, something closer to being tuned out rather than tuned in. (Crawford, ibid., 82)

All the time, this murmur of voices is continuing, whether or not we are tuned to it, a kind of background hum of commentary and conversation, which now and then captures our attention with a pertinent or arresting idea, just as suddenly from the radio, a song, a voice, a sound or a provocative statement will stop us in our tracks and hold our attention:

> The conversational field of activity in these online spaces is dispersed and molecularised, a constant flow of small pieces of information that accrete to form a sense of intimacy and awareness about the patterns of speech, activity and thought across a group of contacts. Like radio, they can circulate in the background: a part of the texture of the everyday (ibid., 82–3).

Also like radio, Twitter and its kin in the field of social media utilise archaic words of transmission in their modern form: (*Instagram* for example, harks back to *telegram*, a form of communication ironically probably unknown to most of its younger users). We send and receive all these reflections of ourselves through *wireless* technology, and the very word, 'Twitter', has sound at the hub of its meaning, 'either as the calling of birds, or the "idle" chatter of humans' (ibid.).

Dear Jean…Love Keith

We have seen that the informal letter can have the capacity to give us the vernacular voice; best among these, as with the Keats letter, is when circumstances do not allow for careful construction and consideration. Some letter writers from the past might almost have had one eye on literary posterity in the process of composition; these are often the least interesting. It is the hurried note that usually gives us the voice without affectation. Where this differs from the instant text message or hurried tweet is that the letter—and in this context we are discussing the letter when written by hand—has a sense of time about it; here is the hand of its author, complete with writing style, conveying a personality even before the words themselves come to our consideration. Letters scribbled during wartime, either in brief breaks in fighting or training, convey the urgent presence of the writer through their necessary informality. In an undated letter, possibly written during the late spring of 1941, the poet Keith Douglas wrote to his girlfriend, Jean Turner, from his training camp in Wickwar, Gloucestershire. Keith and Jean had been friends from university in Oxford: Jean had been at St. Hilda's when he went up to Merton in 1938, and his letters to her from this time have the easy informality of a telephone conversation:

I suppose you can't come this weekend can you: I shan't mind if you don't, except that you'll have further to go next time. We are going on a gunnery course in South Wales for a week and from there straight to Horley or some such place in Surrey. It's possible we may move again from there almost at once. We go on manoeuvres on Monday, arriving back on Friday, go away to Linney Head on Saturday and straight from there to Surrey. However I suppose you couldn't come there. I shall be awfully disappointed if you don't come fairly soon though (if you don't, I expect you won't at all). It would be best of all if I could get the weekend off. Actually it wouldn't be hard to get from Saturday after duty till Sunday night if I stayed in the area.

Incidentally if you careless talk about this there will almost certainly be an invasion and a successful invasion. Love, Keith (Douglas, ed. Graham, 176)

Douglas had a somewhat cavalier personality, a certain arrogance as some described it, which was partly perversely it seems why he was so attractive to women. That quality is suggested here in a kind of 'devil-may-care' attitude to Jean's acceptance—or not—of his invitation. Yet the detail of his movements, which he jokes about in the last sentence (referring to a familiar wartime poster, 'Careless talk costs lives'), suggests an urgency in his need to see her, and a moment of apparent doubt, when he considers she might never come, which may in fact have been placed in the letter as a kind of moral blackmail. All told, this letter, selected almost at random, but quite representative of the tone of his correspondence at the time, gives us a personality that might well have been recognised by his associates, even without a signature. It also conveys eloquently, the mood and atmosphere of the time. Douglas was a tank commander and was killed in France a few days after landing in Normandy in the offensive of 1944.

A voice from relatively recent history, such as that revealed in the letters of Douglas, offers a tone that is quickly recognisable to us; it sounds much as he would talk and has something of the linguistic ring of the present day about it. More surprising, and fascinating, is to hear an informal domestic voice from further back in history, particularly if, as with Keats, its tone seems to run disarmingly at odds with our more familiar perception of an author's more formal style bequeathed to us through their poetic or pro-fessional writings. Great figures of biography—literary or otherwise—are often defined by their oeuvre or, in some cases, the mystery of their hidden lives or the romantic tragedy of their death. To find them gossiping through correspondence adds a charm and unexpected piquancy to their personality that enriches our knowledge, understanding, and even affection of and for them, and gives them revivified life, a three-dimensional quality that might not otherwise be apparent. Here is a twenty-three-year-old Emily Dickinson, writing to her elder brother, Austin, in about 1853. Austin was a lawyer, and taking Emily's letter to him as evidence could be given to jargon and perhaps pedantry:

> I have had some things from you to which I perceive no meaning. They either were very vast, or they didn't meaning anything. I don't know certainly which. What did you mean by a note you sent me day before yesterday? Father asked what you wrote, and I gave it to him to read. He looked very much confused, and finally put on his spectacles, which didn't seem to help him much – I don't think a telescope would have assisted him. I hope you will write to me – I love to hear from you, and now Vinnie is gone I shall feel very lonely…Love for them all if there are those to love and think of me, and more and most to you, from Emily. (Dickinson, ed. Fragos, 218)

We hear the *poetic* voice of Dickinson elsewhere in this book, but here is her *family* voice, hinting through its syntax at a gentle Massachusetts accent, and full of familiarity and a rather impish, mischievous wit, while yet retaining something of the precision of her poems, which might declare itself to the attentive listener to her better known and much anthologised writing. As the editor of Dickinson's letters, Emily Fragos points out, regarding the larger body of the correspondence, 'her letters, with their feverish observations, metaphors, allusions, paradoxes, hyperbole, and rapid leaps of imagination, *must* have confounded their recipients – even if they *were* used to "Emily being Emily"' (Dickinson, ed. Fragos, 8). The letters inform the poems, because they give us a clear imaginative voice to hear when we read them silently to ourselves. It is a strange process, the metamorphosing of a seen text, acting on imagination and memory, into a non-physical sound. Garrett Stewart has expressed this idea perfectly: 'When we read to ourselves, our ear hears nothing. Where we read, however, we listen' (Stewart, 11).

We are thankful that we have these fragments of evidence to convey so eloquently the thinking voice of their authors. Volumes of 'Collected

Letters' by any author may be rarer for future generations of readers and scholars, conclusions may be harder to reach as the email, the tweet and the text message drive development of the culture of instantaneous communication, and we count our interaction with one another in terms of characters, be they abbreviated words or simply symbols. Some of us may disguise our emails as letters through long-term habits, with a traditional salutation at the beginning and end, but we cannot disguise the crucial difference, which is that of time. As Angela Leighton suggests, it is the physical artefact containing the thought, and the temporal progression this fragile paper takes in letter writing, that somehow carries a part of its maker 'as intimate object, addressed to you, signed by me, sealed and sent on a journey' (Leighton, 183).

It is because the letter travels through time and physical space, as do we ourselves, that the process has the potential to give it a meaning that at its most profound, may contain a tinge of melancholy. When we receive a letter, its contents are already out of date, we are reading the past, hearing a voice speaking of a situation and circumstance that was part of their present which no longer exists for them, but which has now been transferred to us. Leighton reminds us that our awareness of this poignant system of time-shift is by no means governed by today's comparisons with instant messaging; it is inherent in the *event* that the letter represents:

> The letter must travel to reach its reader, the message dependent on its delivery. This sense of time taken and space overcome defined the letter long before other modes of communication threw it into relief...Something about the weighty substance of the form is defined by the labour of time taken: the time it takes to write, but also to send, to be received, to be read, pondered, and answered. Between missive and response there is a lag, itself defining the object which must cross it. (ibid.)

We may instinctively feel that, at life's most significant crossroads, the instant email or text may not suffice. There are still times when we feel the need to say something by committing our voice to paper and ink, by-passing even the word processor, in order to truly speak.

Sound itself, then, is both physical and imagined. Using the reality of the physical in memory, sound haunts the mind with the ghost of its voice, and, because the eye is complicit in this process, what we see informs what we hear, either in reality or in the memory. I can conjure sonic images by looking at the words on the page or the screen; likewise I can remember the words, and pull those same sounds back to mind, and even visualise the actual images of the words in my 'mind's eye' as well as its inner ear. This duality is at the heart of understanding the voice or voices transmitted to the imagination as we read: 'If actual sound is itself a transient passenger, invisible and always to be interpreted by the ear, how much more acute is the strange interpretability of sound in the written word, the ghost effects of which are built into its workings. For all reading is a matter of hearing things, in both the literal and the ghostly sense of that phrase' (Leighton, 5). As we read, we hear through the mind, and as it listens, the listening eye of the mind sees, because it has become an imaginative imperative to do so. Seeing through the words of the tweet or the Facebook message, we look for images, either of the subject under discussion, or the actual face of the writer, not as they are represented by the (often idealised) personal photograph attached to the site, but at the moment of writing, which may after all, be this instant or just a few seconds ago. And as we send our own thought into the atmosphere, we have no idea if it will be heard, regarded or responded to. We are like the DX-er in Welles's *War of the Worlds* in an earlier chapter, waiting, hoping for a response. Or the lonely rider in Walter de la Mare's famous poem:

> "Is anybody there?" said the Traveller,
> Knocking on the moonlit door… ('The Listeners', de la Mare, 84)

It is the hope we have for our personal message, sent perhaps in the middle of a sleepless night, to connect, to be heard, the desire to be part of a community that listens and shares across oceans. It is also the power the poet or novelist evokes, when they place us in a sound world, sometimes leaving us there to find our own way home:

> Never the least stir made the listeners,
>   Though every word he spake
> Fell echoing through the shadowiness of the still house
>   From the one man left awake:

Ay, they heard his foot upon the stirrup,
And the sound of iron on stone,
And how the silence surged softly backward,
When the plunging hoofs were gone. (ibid.)

Note

1. It would seem that Haydon's copy was not the edition that inspired the poem, but the Chapman edition was still very much the currency of the time, replacing for many, earlier versions by Dryden and Pope.

10

Full of Noises: The Sound Map of Self

The Body as Communications Hub

The most salutary place to begin listening to existence is in an environment where seemingly it has no place, in an anechoic chamber, a space with walls that are so echo-free as to completely absorb reflections of sound. The term 'anechoic' was coined by the American acoustics expert, Leo Beranek. Technical measurement of sound in such places would show a reading of *negative* decibels. To spend time, even ten minutes, in one of these studios is a strange, and for some, an almost intolerable experience. Speak, and the room seems to swallow the sound; turning and walking, even a few steps, becomes problematic, because the perceptible sound cues the world gives us to enable ambulatory abilities have been removed. Curiously, however, seated in the centre of an anechoic room, as the ears and mind adjust to this eerie auditory vacuum, one starts, after all, to somehow hear things. This can be perplexing; it is often only afterwards, during debriefing sessions, that the reality of the experience becomes clear. In quiet spaces, when we are in bodily repose, we hear more. In a room with no sound, we encounter the ultimate listening experience: ourselves, the beat of the heart, the soft whine of our nervous system, the low rumble of our blood

© The Author(s) 2019
S. Street, *The Sound inside the Silence*, Palgrave Studies in Sound,
https://doi.org/10.1007/978-981-13-8449-3_10

stream. In an anechoic chamber, we become sound, and by so doing, we prove the reality of our existence and the fundamentals of our individual identity. We are each of us vibrating at infinitely various frequencies, and our whole body is listening for evidence of the world around us. Even when sound apparently stops, we experience it through vibrations that may sometimes be even beyond the capacity of our ears to hear, absorbed by bone conduction. As a microcosm of the sonic world, we start within our own being, hearing ourselves when we eat or drink, when we scratch our head or yawn. We prove by such simple acts that our ears are by no means the only sound receivers at our disposal. When we speak, we hear ourselves internally through a cave of bone, flesh and blood; little wonder our voice seems so strange when we first hear it played back to us. Modern audio technologies such as non-invasive headphones take advantage of this natural phenomenon through the use of bone conduction transducers that guide micro-vibrations through the cheekbones to the inner ears. By so doing, they conduct sound without plugging or covering the ears, enabling them to get on with the task of delivering the ongoing signals of our surrounding environment independently of our selected source of audio. It is a powerful demonstration of how much our physical frame plays its part in the absorption of several sonic originations at the same time. When we walk, we move to the iambics of the human heart, which may perhaps be why walking can take us on a journey of self-discovery, as we move through landscapes and cityscapes with all our senses conjoined in an exploration of the world as context for who we are at a precise moment of being.

This then is where it begins and ends: in the silence that is not silence, the sound of ourselves with its rhythmic murmurs, the vase of sonic being that is the receptacle for the music of the world that floods in to fill it, the apparently empty field which is an event in itself, altered by the playing of the first sound. Outside, in another, physical field, a dog is barking, a flock of birds fly over, wings beating like a round of applause; the wind is getting up now, and the event of it changes the mood again. It is also our mood, because we are here to hear it change. As we listen, memory and imagination suggest a history to the place, the children who play here at the weekend, the walkers, the changing seasons reflected in the growth of the grasses and trees, the sound of the gate that closed behind me as I entered. If I wrote music, I might find a song in it, a series of notes that could link me to the place, to create a pattern of sound that could only come from here, the field and myself, in partnership. There might be a whole radio programme to be made from just these things, a soundscape that through actuality, words and music would transmit the essence of the place, would make it at one and the same time localised and universal. This field belongs to everyone. I just happen to be here now. It need not even be a field; it can be anywhere where we pause and tune our attentiveness. Several times during this journey, we have found our path converging with the sound world of the writer Lucy M. Boston. In her novel, *The River at Green Knowe*, Ida suggests to her friends Ping and Oskar, 'Let's shut our eyes and say everything we can hear' as they sit in their canoe in the still morning:

> "Water under the canoe's ribs, whirlpool round my paddle, drip off the end of Ping's paddle, bird flying off tree, lark singing, rooks circling, swallows diving, rustling in grass, grasshoppers, honeybees, flies, frogs, bubbles rising, a weir somewhere, tails swishing, cow patting, aeroplanes, a fishing rod playing out; zizz, buzz, trill, crick, whizz, plop, flutter, splash; and all the time everywhere whisper, whisper, whisper, lap, chuckle and sigh." If someone moved in the canoe, a moment later on the far side of the river all the rushes nudged each other and whispered about the ripple that had arrived. "Everything's trying to say something," said Ping. (Boston, 216)

Before they began listening, and after, when their attention was diverted by conversation or the dynamic of the visible world, these sounds existed; through Boston's black words on white paper, they are held, and the moment is recorded, but the experience has happened, and moved on. It is as the old philosophical saw asks: if a tree falls in a forest and no one is around to hear it, does it make a sound? Like Ida's river, my field has been going on without me, I have walked in on its continuing performance of itself, as if entering a concert hall half way through a symphony. For a moment, I adjust to the presences around me, and then I become a resident in the space with everyone else, part of what is happening here. 'Events then', wrote Roy Bhaskar, 'are categorically independent of experiences'.

There could be a world of events without experiences. Such events would constitute *actualities* unperceived and, in the absence of men, unperceivable. There is no reason why, given the possibility of a world without perceptions, which is presupposed by the intelligibility of actual scientific perceptions, there should not be events in a world containing perceptions which are unperceived and, given our current or permanent capacities, unperceivable. And of such events theoretical knowledge may or may not be possessed, and may or may not be achievable. Clearly if at some particular time I have no knowledge of an unperceived or unperceivable event, I cannot say that such an event occurred…But that in itself is no reason for saying that such an occurrence is impossible… (Bhaskar, 32)

Although the sounds of Lucy Boston's river is an event experienced silently in my mind as I read her account, the reality is also based to a degree on a personal memory, because in fact I have visited Hemingford Grey, where the original house that was the model for Green Knowe stands, the oldest continuously occupied family home in England. By imaginatively 're-hearing' the sound of the river, preserved in my mind, I can mentally travel there, and yet I am aware that its physical existence is independent of my presence. Likewise, when I listen to Wifi radio, I tune regularly to one of my favourite music stations, Jazz24 from Seattle. Yet if the internet connection goes down and other ambiences flood back into my space, Jazz24 will continue to play its music for the rest of the world until my signal is restored and I rejoin the audience. To repeat Bhaskar's phrase, 'events are categorically independent of experiences', so it is our experience of them that proves our reality, more than theirs. Even when we have no means to experience them, we have the internal capacity to make an imaginative event that creates an internal experience, especially if it is based on memory.

The realisation that there is a self and a non-self comes upon us gradually as children. Beginning with the sound world of a mother's womb, we are then flooded with light and the roar of air and all it contains, like a swimmer coming to the surface. We discover the world, piece by piece, serendipitously, and among the first things, we come to recognise are voices. Self-awareness dawns when we first make distinctions between the sensations we receive from the world, and the conditions that have pertained to produce those sensations. A lullaby sung by a mother to her child fascinates and lulls because it comes from a loved and trusted voice, modulating itself and weaving into the mind. Our first existence is as receivers of signals. When the physical receivers fail us, be it technology or our human frame, the ghost of the sound ignites or continues inside the mind. Moving beyond the experience of the event, we can invoke its imaginative presence, but in the end, the event will happen or continue without us. The great prose poet of the natural world, Richard Jefferies, was so intimate an observer of the flora and fauna around him in the Wiltshire countryside that by the time of his premature death at the age of 39, he felt himself to have moved *beyond* observation; in *Hours of Spring*, possibly the last essay written by his own hand, he utters a passionate and heartbroken cry:

> I think of the drift of time, and I see the apple bloom coming and the blue veronica in the grass. A thousand thousand buds and leaves and flowers and blades of grass, things to note day by day, increasing so rapidly that no pencil can put them down and no book hold them, not even to number them – and how to write the thoughts they give? All these without me – how can they manage without me? …But today I have to listen to the lark's song – not out of doors with him, but through the window pane, and the bullfinch carries the rootlet fibre to his nest without me. They manage without me very well; they know their times and seasons…They go on without me. (Jefferies, 22–3)

Dream Voices

All of which raises the question, am *I* still there when my senses become dormant? Do I continue when my brain switches off? As I sleep, I dream, and the unconscious takes over, and Susan Greenfield suggests that 'consciousness could be seen as the actual, first-hand experience of a certain mind, a personalised brain. Consciousness brings the mind alive…It is your most private place' (Greenfield, 149). Yet here we really do enter the shadow worlds of conjecture; when I dream, I am sometimes disturbed by the sounds of the world, although they may not always be enough to wake me, to switch on my consciousness. These sounds may enter the narrative of my dream, just as I entered the event of the field, contributing to it. Sometimes, while dreaming, I hear myself thinking, 'this is a dream'. The sound that impinges from the physical world may have the capacity to change that narrative or build on it. In truth, all my systems will never fully switch off until I die, until the power is terminated. So we may say that in fact, dreaming is not a factor of unconsciousness, but is a further facet of *consciousness*. It is also the place to which the journey upon which we embarked in our search for silence ultimately takes us, because we now enter the world of sonic imagination, not as an observer, but as a participant. Here, Richard Jefferies *can* be part of the natural world from which his imminent departure seemed so final, because the mind and the body in partnership employ memory and the surrounding sounds of the ongoing world to create a continuum through sleep that allows us to continue the story tomorrow. Taking a coffee break from my work, I have the choice of either switching off the computer completely, or putting it into 'sleep' mode. So it is…Even as I drink my coffee, staring out of the window at the field at the bottom of the garden, I invoke a kind of dream state of reverie. I am 'day-dreaming', apparently thinking of nothing, until out of that nothing comes the understanding that a sound or a new thought has occurred.

Sleeping, my dreams and the sounds they contain may be affected by mental/physical issues 'as a result of certain states of brain activity that cannot process large amounts of sensory input because we are asleep' (ibid., 59). If an anxiety from the day creeps in and begins its lisping, or if the memory of an outstanding task or unfulfilled idea nags, or a recollection of a shouting match during a disagreement starts a replay, the dreamscape may try to deal with it, or if it cannot, I may awake, to enable the brain to decide on a course of action. Through all this, there is interactive imaginative sound. Even my diet or chemistry may be a factor. Did I eat or drink too much last evening? Did I take too many painkillers…or perhaps not enough, because something is hurting, for sure. Or it may be something ongoing about a mental state that needs attention. Dreaming can become very real for many people, and that border shadow-land can create confusion as to which reality is the dominant force within us, 'as in schizophrenia, [when] prevailing chemicals have limited the efficiency of large-scale dialogues over large banks of brain cells' (ibid.).

There is a hinterland between waking and sleeping, a state of drowsing, when the mind enters a relaxed reverie. In this borderland of consciousness, we may find ourselves inspired by a still small voice offering ideas, images and insights that seem to hint at unexpected levels of inspiration and creativity. Vowing to remember these pearls of wisdom the next morning, we are often defeated; the writing on our mental blackboard has been too faint to last, and the night has wiped the slate clean. It is for this reason that many keep a notebook at the bedside, to provide a prompt for recovery of the thought. Emily Dickinson expressed it perfectly in a short poem written in about 1861:

> I held a Jewel in my fingers –
> And I went to sleep –
> The day was warm, and winds were prosy –
> I said "T'will keep" –
> I woke – and chid my honest fingers,
> The Gem was gone –
> And now, an Amethyst remembrance
> Is all I own – (Dickinson, 112)

Here is a mystery; sleep takes us to places in which we are ourselves and yet transports us to a fundamentally different place in which the rules of consciousness seem to work in quiet conflict with another layer of being. It can take away those jewels of last thoughts like a temporary death, perhaps prior to the night workers who arrive to prepare the office inside the mind for another day. At the same time, it gives us dreams, some of which we retain, while some vanish into the light like Emily's poem. Some composers believe that the dreamt sounds of music can be recalled in their entirety by the waking mind. Berlioz claimed to have retained the dream of the first movement of his *Symphonie Fantastique* completely the next morning. It may be, as Massey suggests:

> Music is the only faculty that is not altered by the dream environment, whereas action, character, visual elements and language may all be modified or distorted in dreams…Music in dream does not become fragmented, chaotic or incoherent, neither does it decay as rapidly as do the other components of dreams on our awakening…What distinguishes music

in dreams…is that it is consistently normal…One might say that music never sleeps…It is as if it were an autonomous system, indifferent to our consciousness or lack of it. (Massey, 42–50)

The dream state may bring gifts, but it does so under its own terms; this is a place in which, while we continue to be ourselves, we seem to be manipulated by another will and another set of values. 'The night dream (*rêve*) does not belong to us. It is not our possession. With regard to us, it is an abductor, the most disconcerting of abductors; it abducts our being from us…we become elusive to ourselves…' (Bachelard, 145).

We know that there is a long history of artists, composers and poets seeking to attain the creative dream state through artificial or conscious means, be it alcohol, drugs or meditation. Samuel Taylor Coleridge subtitled his great poetic fragment, 'Kubla Kahn', written in 1797, 'A Vision in a Dream'. The dream, in fact, was induced by the consumption of opium, after reading a text about the palace of the Mongol emperor of China, Kublai Khan: Xanadu. On waking, he wrote the beginning of what might well have been a much longer piece, created in a white-hot burst of creative mnemonic energy. He had the whole work in his head: it was as though he was taking dictation. Only when he returned to his desk after being interrupted by a visitor at the door, did he realise that 'the Gem was gone', in Emily Dickinson's words. Yet what survives contains astonishing imagery, powered by the sheer sound of words:

> The shadow of the dome of pleasure
> Floated midway on the waves;
> Where was heard the mingled measure
> From the fountain and the caves.
> It was a miracle of rare device,
> A sunny pleasure-dome with caves of ice ! (Coleridge, 118)

Coleridge may have considered that the 'voices' of his dream, had more to tell him, but the fragment ends with lines that give the work a finality of its own, and could even be read as a metaphor for the fragile state of dream consciousness:

> For he on honey-dew hath fed,

And drunk the milk of Paradise. (ibid.)

Whether it be music or voices, benign or malign, the mind in the dream state seems to enter a subjective reality than can have the capacity to inform the personality and transform thought processes; John Keats begins his 'Ode on Melancholy' with what almost seems like a rebuke to Coleridge's methods:

No, no, go not to Lethe, neither twist
Wolf's-bane, tight-rooted for its poisonous wine…

For Keats, it is the dream state without the support of external stimulants—waking or sleeping—that has the power to dictate creative terms, and the best thing we can do is to wait for it. This is the phenomenon to which he gives the name, 'Melancholy', and it is in this very condition, the ephemeral fleeting world of dream, be it waking or otherwise, that its very imaginative potency exists; he expresses the idea with consummate precision in the last stanza of the poem:

She [Melancholy] dwells with Beauty—Beauty that must die;
And Joy, whose hand is ever at his lips
Bidding adieu; and aching Pleasure nigh,
Turning to poison while the bee-mouth sips:
Ay, in the very temple of Delight
Veil'd Melancholy has her sovran shrine,
Though seen of none save him whose strenuous tongue
Can burst Joy's grape against his palate fine;
His soul shalt taste the sadness of her might,
And be among her cloudy trophies hung. (Keats, 219–20)

We dream in the world we inhabit, according to the context of when and where we are. In Chapter 4, we considered a sense of place and its effect on the composers and musicians who inhabited various terrains. Yet we must remember time as well as place; Mozart, Beethoven and Haydn fell asleep each night thinking of the sounds and ideas with which they had been surrounded during the day, and it was the culture, attitudes and pace of life they witnessed that must have coloured their dreams, just as it shaped their music. Many years ago, I had a conversation with the Austrian-born conductor, Rudolf Schwarz. Growing up in Vienna before and after the First World War, Schwarz was a pupil of Richard Strauss, and he talked to me of Strauss's teachings, in particular on the subject of relative tempi in music and conducting: 'Music is written in its time, and times and the pace of life changes. When Mozart wrote "Presto" (fast) on his scores, he was thinking in eighteenth century terms, and the fastest thing known at that time was the galloping horse. That was his reference point. Today we have powerful cars, jet aircraft and space rockets. Naturally we find ourselves drawn to play the music of the past faster than it was intended in its day'. Our dreams may take us to other states of consciousness, but we always awake in the same place and time, and life goes on.

Sleep is of course about more than dreaming, because during sleep the brain stockpiles chemicals to enable us to function in our conscious world. It only takes a few days of being sleep-deprived before these interactive functions start to decline. Why does the brain 'know' when to fall asleep? Answer (partly) because the light-sensitive pineal gland can read lightness and dark and can feed the awakening senses with doses of melatonin. We break our silence as we break our fast: 'Good morning' we say. The cockerel crows, and the field of consciousness awakens for another day.

The Cross-Referencing Mind

Listening now to the field on the bottom of my garden, I become aware of the sound under the sound that I had at first thought of as silence. There is the ambience of the surrounding area, gradually grading itself into the perspective of distance. Now I can hear traffic on the road beyond the trees, and beyond that, the noises from the container port on the river. It turns the imagination back beyond the moment, to previous moments, to what this place might have sounded like had I been standing here one or two centuries ago. History books can give me some sort of a context, but there comes a point where I have to imagine what the ghosts heard here, or how it would sound if a close encounter of the third kind occurred in front of my ears. The field becomes other fields, transforms itself into a place of memory, a real field, but not the one I am standing in now; this is where I walked with my father when I was a boy. Now I hear his voice and the bark of my grandmother's dog, demanding a ball be thrown. I throw it high in the air, and the dog chases it, but when it lands, dog and man have gone. My daydreaming is suddenly interrupted by a shout from the mobile telephone in my pocket; these things are always 'urgent' and so I attend to it. It is a text message, written in capital letters, which means of course that this is indeed a matter to be attended to immediately. I am late for a meeting. I turn from the field, and return to the dynamics of UK time, but my mind has logged the imaginative journey, and memory has archived it in the compartment devoted to individual and personal experience, the unique and only mind that is myself, listening as it does to the world from a perspective belonging to no other. I think of Henry David Thoreau, writing in his journal in August 1852:

> I only know myself as a human entity, the scene, so to speak, of thoughts and affections, and am sensible of a certain doubleness by which I can stand as remote from myself as from another. However intense my experience, I am conscious of the presence and criticism of a part of me which, as it were, is not a part of me, but spectator, sharing no experience, but taking note of it, and that is no more I than it is you. When the play – it may be the tragedy of life – is over, the spectator goes his way. It was a kind of fiction, a work of the imagination only, so far as he was concerned. (Thoreau 2009, 163)

A person alone in a room, apparently silent, is nonetheless experiencing an imaginative soundtrack made up of thought, memory and emotion. The triggers that set off this emotional sound inside us are often today given expression through social media, responding to a surrounding world which, although it may come to our door through print or the visual interpretation of feelings and attitudes, nonetheless is full of emotional sound. As we have seen in the previous chapters, the emoticons, the screaming of headlines, the sudden noise of upper case letters to drive home a point, all conspire to bludgeon and bully thought by using instant messaging to bellow opinion and dogma around the world within an instant. The new social media, its causes and effects, loneliness and community, news, fake news and the isolation of the individual offer those of us who may have no other way of interacting with the world, the experience, and in some cases, the impression of being heard and making a difference. They may provide disenfranchised voices with the illusion of being 'heard'. The effect of this on fragile mental states can be devastating or stimulating. The internal sounds within the mind that dictate and justify much of what we do often depend on the nature of those 'heard' voices, as in mental illness and the interconnectedness of sensual stimuli.

In Chapter 8, we discussed Kandinsky's theories of the relationship between colour and sound; this is a complex issue, and caution is required when seeking to assign such a gift to a neural condition such as synaesthesia. An in-depth study of the condition is outside the scope of this book, and anyone interested in pursuing the subject further may wish to study the work of specialists in the field.[1] There are specific factors that affect the relationship between colour and sound; a number of artists and writers who believed themselves to be synaesthetes may in fact have been experiencing other conditions. Among them, we may consider briefly the case of the French poet, Baudelaire, whose poem, 'Correspondances', translated as 'Connections', supports his assertion that for him, 'sounds are clad in colour' (Baudelaire, quoted by Harrison, 116).

> Like long echoes which from a distance mingle
> Into a shadowy and deep unity
> As vast as night and light
> Perfumes, colours and sounds reply to one another. (ibid.)

Vivid as this is, John Harrison, in his book on synaesthesia, casts doubt on the idea that Baudelaire was in fact naturally synaesthetic, considering other issues that may have been contributory to his experiences, including physical health and the use of artificial mental stimulants:

> Baudelaire suffered from syphilis, a disease that in the later stages can attack the brain. We also know that a number of psychoactive substances can give rise to synaesthesia…A contemporary of Baudelaire's, the French scientist Gautier, reported in 1843 that he had been able to produce "pseudo sensations of colour" artificially, in particular by the use of hashish. Other drugs, such as LSD, mescaline…and psilocybin, so-called "magic mushrooms", have been reported to cause confusion between sensory modalities. In some

cases, use of these hallucinogens has caused sounds to be perceived as visions. (ibid., 118–9)

Our personal island is indeed 'full of noises'; we create our own sound-track, whether we induce it or endure it, and even in sleep our dreams are full of strange music. What are these sounds? An earworm that stays all day? A remembered voice that haunts our waking hours? We are indeed 'such stuff as dreams are made on'. That said, they are *our* dreams in *our* mind, peopled by *our* sounds, and the mind is, as David Fontana has said, 'a direct, lived experience…To experience one's own mind is to experience the strange mystery of life itself, something that can be known only from the inside' (Fontana in Lorimer, 72). Yet other existences impinge and shape that experience as the sonic world murmurs or shouts around us. Everything is interactive, and we are at the centre of a complex dialogue of sensory events that exists independent of ourselves, while happening for us subjectively as we move through it. 'Every time we listen, not just to music, but to anything at all, our auditory perception is the result of a long chain of diverse and fascinating processes and phenomena that unfold within the sound sources themselves, in the air that surrounds us, in our ears, and, most of all, in our brains' (Schnupp, Nelken and King, viii).

Concerning the mystery of conjuring art from the partnership of mind and body, the poet Dannie Abse, who was also a doctor, once wrote: 'I do not know how to write a poem. If I did, I would be able to write one at will… Poetry is written in the brain but the brain is bathed in blood' (Abse, 184). The sounds we hear belong to us when they touch us, but our relationship with them, as with all phenomena, is subjective/objective. Brain, mind or both? However we define our response-machine, it is unique, as Amit Goswami, professor of physics at the University of Oregon reminds us, 'a place where the self-reference of the entire universe happens. *The universe is self-aware through us.* In us, the universe cuts itself into two – into subject and object' (Goswami, 190). We experience something for the first time, and we may respond instantly or over a period of time. A sound startles us, and our heart rate rises. The sound may then pass into memory and be reflected upon later. It is memory that comes to our aid as a crucial interpreter of the world. In order for Frederick Delius to write his tone poem, *On Hearing the First Cuckoo in Spring*, he must first have *heard* a cuckoo before re-imagining it as music. Music is made in the present moment, but the memory of sound preserves us from the fate of living in that moment as a *permanent* present, because as we have already discussed, as mortal and temporary inhabitants of our space in the world, we share sound with time, which is also fading to silence as it happens around us. In his *De Memoria*, Aristotle muses on the nature and role of human memory thus:

> When memory first has been produced in the individual and ultimate organ of sensation, the experience and the knowledge in question…have already existence in the experiencing subject. But memory in the proper sense will not exist till after the lapse of time. We remember in present time what we have previously seen or heard, we do not now remember what we have now experienced. (Aristotle, 109)

We are echo chambers, and the memory of significant sound rings in a curious way, because as it fades into time, it may magnify imaginatively insignificance within the mind. A childhood recollection of a song or the voice of a long lost loved one: internal listening amplifies these things. All this within a world that grows more claustrophobic and more noise by the day. Global transport has brought the horizon closer, while 'at the same time…voices clamouring within that horizon have increased exponentially' (Smith, B., 341). It becomes harder to listen with due attention to the sound that matters. 'In those circumstances two courses of action present themselves most insistently: to speak very loudly oneself or to shut one's ears. Ecological viability lies in a third possibility: to look, to listen, and to know the difference' (ibid.). Listening to what the imagination plays us, either in terms of memory of immediate perception, is a way of seeing, as the bell rung at the beginning of this book would remind us. By starting with its clamour and following its journey until its ring merges with the air around it, we are, as Keats said when we commenced our journey, brought back to our solitary selves.

Homecoming

It is in our selves, after all, that the answers to the questions, as well as the questions themselves, are to be found. It is all a matter of tuning, and generating a potential for listening and understanding, or, to use media analogies, to switch from mono to stereo or from monochrome to colour. About 130 pages into John Steinbeck's novel, *East of Eden*, at the beginning of Chapter 13 to be precise, there is an extraordinary moment when the author suddenly breaks into his own narrative and philosophises. On first reading, it takes us by surprise, and briefly we may wonder at his reason for this personal soliloquy. It is, of course, relevant to Steinbeck's larger unfolding story, just as it is relevant at this point in ours. 'Sometimes, he writes, a kind of glory lights up the mind of a man…'

> …It happens to nearly everyone. You can feel it growing or preparing like a fuse burning toward dynamite. It is a feeling in the stomach, a delight of the nerves, of the forearms. The skin tastes the air, and every deep-drawn breath is sweet…It flashes in the brain and the whole world glows outside your eyes. A man may have lived his whole life in the gray, and the land and trees of him dark and sombre…And then – the glory – so that a cricket song sweetens his ears, the smell of the earth rises chanting to his nose, and dappling light under a tree blesses his eyes. (Steinbeck, 133)

Steinbeck's 'glory' might be interpreted as 'transcendence' but whatever the word, it can result in the ability to heighten the senses to such a level of highly tuned sensitivity that it can sometimes be the trigger for experiences of almost metaphysical impact. It is, we might suggest, a kind of concentrated empathy, and what could be more sonically empathetic than to develop the gift of engaging through active listening to surrounding and internal sound, absorb its meaning, and attend to what might happen next? After all, silence in the living world may be, as we have said, an illusion, as a new-born sound replaces the dying ember of the last. Notwithstanding, there may be pauses, spaces, brief windows within the continuum of experience into which the sounds of life pour, enabling new responses to emotional, cultural and spiritual stimuli. Every person our own sound studio, we each of us possess the continuing capacity for listening to ourselves, even when external forces lapse into apparent stillness. These sounds may be positive, or they may be destructive and dangerous, as in forms of mental illness and psychological degeneration. We may underestimate the power of sound—in particular the voice—in even the simplest sonic event, to trigger new directions of thought, and refresh ways of responding towards receptivity for new adventures.

I am conscious that I am writing these words, and that some sort of internal voice is speaking them into my fingers as I type. You are reading them, and your imaginative voice is interpreting them. I am speaking to you, or is it that you are a musician, playing and interpreting the notes of meaning on the instrument that is your mind? At the end of this journey into the sonic imagination, we are confronted with the self, and thus, with more questions:

> Is it *you* speaking to *you*, or are *you* the thing that is endlessly spun by that conversation? In which case, where do you go when the voice stops? Does it ever stop? Who is the "me" or "you" to whom a young child speaks aloud, and who is the speaker – especially at the stage where the fragile self is still in the process of being formed? Who speaks to the novelist in her study, or to the psychiatric patient in his hospital room? To the churchgoer, praying silently in her pew, or to the ordinary voice-hearer, listening into the transmissions of a fractured self? (Ferneyhough, 258)

The French philosopher and phenomenologist Maurice Merleau-Ponty suggests that the event of a spoken (or written) word, emitting from us into the world, is part of our physical presence:

> I reach back for the word as my hand reaches towards the part of my body which is being pricked; the word has a certain location in my linguistic world, and is part of my equipment. I have only one means of representing it, which is uttering it, just as the artist has only one means of representing the work on which he is engaged: by doing it. (Merleau-Ponty, 210)

Yet we would argue that before the word is uttered physically, it must be conceived, 'heard' internally, which returns us to Abse's statement of existential duality. It is clear that our bodies are the instruments upon which our mind—memory, inspiration, conception and expression—plays. The bodily function of speaking connects us socially to the world, and our need to communicate lies at the root of every act. These words are language, on a printed page, but they are more than that, because at the basis of their being is an utterance, and an utterance is a physical unit of social behaviour, seeking to express something to another human being. They may be hieroglyphs, but these words are also are identification marks denoting sounds, striving to convey meanings. David Jones, in the preface to his long poem, *The Anathemata*, was emphatic that the work was intended to be spoken and heard. The text, in his view, was a means of reading towards sonic interpretation, just as a musician reads a score: 'While punctuation marks, breaks of line, lengths of line, groupings or words or sentences and variations of spacing are visual contrivances, they have here aural and oral intention. You can't get the intended meaning unless you hear the sound, and you can't get the sound unless you observe the score' (Jones, 35). We make music of what we find to help us make sense of our present and help us towards a future, just as David Jones did in his great poem. The score comes to life when it is given expression, and that expression will differ, according to the instrument that interprets it, based on (because here the instrument is human) experience and memory. There is only guidance in the black marks placed so carefully on the page. Yet if we can imaginatively engage sufficiently with their shape and meaning, including the white space around them, they may encourage us to 'sing' them, even within ourselves. It is all music, and 'our ability to make sense of music depends on experience, and on neural structures that can learn and modify themselves with each new song we hear, and with each new listening to an old song' (Levitin, 108). It is an attempt at establishing a communicative conversation with another, based on a desire to come to an understanding of who and where we are in the world, and for this, we use memory and experience as compasses towards an idea of where the journey goes next. Fundamentally, as David Jones said, 'one is trying to make a shape out of the very things of which one is oneself made' (Jones, 10).

A conversation is a sequence of behaviour, expressed through sound, interacting and sharing, and drawing in numerous subtle ways on a whole library of experience and imaginative probing that is stored in the mind. Memory is the contextual engine of our whole communications hub: 'The part played by the body in memory is comprehensible only if memory is not only the constituting consciousness of the past, but an effort to reopen time on the basis of the implications contained in the present, and if the body, as our permanent means of "taking up attitudes" and thus constructing pseudo-presents, is the medium of our communication with time as well as with space' (ibid.). By way of illustrating this, Merleau-Ponty points us to a passage by Proust, in *Swann's Way*:

> When I awoke like this, and my mind struggled in an unsuccessful attempt to discover where I was, everything would be moving round me through the darkness, things, places, years. My body, still too heavy with sleep to move, would make an effort to construe the form which its tiredness took as an orientation of its various members, so as to induce from that where the wall lay and the furniture stood, to piece together and so give a name to the house in which it must be living. Its memory, the composite memory of its ribs, knees, and shoulder-blades offered it a whole series of rooms in which it had at one time or another slept, while the unseen walls kept changing, shaping themselves to the shape of each successive room that it remembered, whirling madly through the darkness…My body, the side upon which I was lying, loyally preserved from the past an impression which my mind should never have forgotten, brought back before my eyes the glimmering flame of the night-light in its bowl of Bohemian glass, shaped like an urn and hung by chains from the ceiling, and the chimney-piece of Sienna marble in my bedroom at Combray, in my great-grandmother's house, in those far distant days which, at the moment of waking, seemed present without being clearly defined. (Proust, 5–6)

Proust's narrator could equally have found the image opening the past here to be a sonic one, the ears after all representing the faculty of auditory reception of the world's presence and his presence in it. But what if consciousness, like energy, is an existing force, like sound waves, that flow constantly, and are caught by the net of the brain? Consciousness works through the brain, and the brain earths it, while storing the elements that it can interpret in various ways, depending upon its state of health and wellbeing. The voice that makes the poem happen may be inspirational for all sorts of reasons, good and bad. With the development of functional magnetic resonance imaging (fMRI,) scientists came closer to understanding inner speech, 'inspiration' and also hallucinatory messages and signals, detecting blood flow in the brain, enabling possible clues to investigate neural activations. What it is less reliable at telling researchers is *when* this activation happens, so it will not help the poet to predict when the next poem will occur, in other words, the inner sound 'event' that will transmit itself ultimately to the brain of the reader. Likewise, the hallucinatory 'voice' that triggers a psychotic episode *is* going to happen, and we know in many cases what triggers it, but the timing of the 'event' of it may remain unpredictable. As the old joke against public transport goes, we know that the next bus is on its way, but there is no timetable, and sometimes we can wait for a long time, before several arrive all at once.

In his novel, *The Opposing Shore* (*Le Rivage des Syrtes*), Julien Gracq gives us a poetic expression of the relationship between the material and sensory worlds:

> …When upon waking we hear, during a long aimless stroll, the resonance of a deeper note…Something like a faraway warning…quicker with admonitions than any dream; perhaps it is the noise of a single footstep on the paving stones, or the first cry of a bird faintly heard in our dissolving sleep; but that echo of footsteps wakens a cathedral reverberation in the soul, that cry resonates as if across the broadest beaches, and our ear strains through the silence toward a void in ourselves which suddenly has no more echo than the sea. Our soul has been purged of its murmurs, of the uproar that dwells there; a fundamental note sounds and rejoices in the newly wakened capacity, which is no more than its own… [It is]…a resonance that surprises us like the footsteps of someone walking through a grotto, rousing echoes…and now we must live…as though in a familiar room whose door swings ajar, suddenly, upon a cavern. (Gracq, 96)

Where some speak of the deepest imagination, others may whisper of the soul; at our centre is a mystery which is none other than our very self. 'In the theatre of consciousness, the natural relationality of the vocal — the acoustic relationality that speech itself, insofar as it is sonorous, confirms – is pre-emptively neutralised in favour of a silent and internal voice that produces a self-referential type of relation, an ego-logical relation between the self and itself' (Cavarero, 46). Here, Adriana Cavarero moves us towards metaphysics, reminding us of Plato's suggestion that 'the soul…can do without the bodily *phone* and contents itself with a metaphorical voice. And from this point on, the soul obstinately speaks with a voice that does not reverberate. When its interior discourse comes out of the mouth and is vocalised, it thus finds itself confronted with a verbal interlocution that spoils the mute and disembodied perfection of the solipsistic colloquium' (ibid.). Any creative artist, in whatever field of endeavour, will recognise the danger of voicing the intention behind an imaginative idea before it is realised.

We have returned to our beginning, to the silence inside that awaits fulfilment, and at the same time, Gracq's beautiful passage takes us deep into ourselves, towards where our journey concludes. Indeed, we might argue that the whole of this exploration has taken place within ourselves, as we respond in our own very deeply personal way, to the stimuli of the world around us. It is a constant dialogue that becomes a monologue, and the difference between them is the difference between movement and stasis. Everything in the physical world is in a state of movement, and sound is no exception, because sound waves travel through the air:

> From the subtleties of body language to the precision of the spoken word to the unambiguity of a simple hug, virtually all communication relies on movement. However global or imperceptible, all [animal] movement depends on the contraction of some muscle group somewhere in the body. If contraction of all muscle is defunct, all that is left is the ability to drool or shed tears (Greenfield, 32).

The word 'animal' derives from the Latin, *animus*, meaning 'consciousness'. We *are* animals, conscious of who and where we are, and always in the process of trying to make a meaning from time and place; we invoke possibilities, draw conclusions based upon evidence, based in turn on what our senses tell us. What we do not know, we imagine. The sound of a bell fading, and the gradual encroachment of ambience around its dying sound, ensures that there is no silence, only an inability to hear. Our attention spans are diminishing. Sound can try our patience because our senses determine that we must experience it in real time, and these days we may feel that to be time we cannot spare. We have to stay with it from start to finish, but even as we register it, it has moved on; notwithstanding, by the time it has happened, it has already become part of us and perhaps changed us forever. Meantime, we continue, each of us, onwards within our own sonic space, our field, waiting for the next sound to alter the imagination. Of the creative impulse, the poet and musician Patti Smith asked rhetorically:

> Why is one compelled to write? To set oneself apart, cocooned, rapt in solitude, despite the wants of others. Virginia Woolf had her room. Proust had shuttered windows. Marguerite Duras her muted house. Dylan Thomas his modest shed. All seeking an emptiness to imbue with words. The words that will penetrate virgin territory, crack unclaimed combinations, articulate the infinite. (Smith, 87)

We could say the same of painters, composers, scientists, philosophers: anyone, you, me, any sentient being has the capacity to find a space in which to wait for the sounds of the imagination. It does not need to be a room, or a hut or a field; we carry our own creative space, our point of departure inside us wherever we are. Patti Smith's space may be her New York desk or a street corner café in Paris. The important thing is to furnish that space with stillness into which consciousness of sound will flow like mercury, but time must be made for the moment under the moment to happen, for the sounds beneath the silence to emerge into the light. W. B. Yeats speaks of 'a ghostly voice, an unvariable possibility, an unconscious norm' (Yeats, 524). Perhaps what has been underlying all these journeys into the sonic imagination is a single rather fanciful idea: that sound, as it passes us, contains the potential to touch a responding note that lies deep within us, perhaps so deep that we only become aware of it even subliminally when certain tones in the air touch it sympathetically, as the wind in a particular direction brushes the strings of an aeolian harp, and a spontaneous reaction occurs and enters our awareness. Once heard and acknowledged, a door is pushed gently open. Yeats again:

> What moves me and my hearer is a vivid speech that has no laws except that it must not exorcise the ghostly voice. I am awake and asleep, at my moment of revelation, self-possessed in self-surrender; there is no rhyme, no echo of the beaten drum, the dancing foot, that would overset my balance. (ibid.)

This may be why we find ourselves quite inexplicably moved by a note, a phrase, a tone or a voice, as it crosses the threshold of consciousness and becomes part of the sound inside the silence. It is not the work of the intellect: in fact, it is something beyond thought, as the pattern of song from a bird's throat touches us before we even identify it, as it surprises out of the air. The beginning and ending lie in the gift and the learnt skill of not anticipating, but allowing consciousness to be visited. The readiness is all. It is only a matter of tuning. The subtle songs that begin our journeys of discovery can be shy; they may lie hidden under other more extrovert sounds, but the mind has the almost miraculous power of extracting them and placing them onto a white canvas, making a mark on the imagination like the strokes of a Chinese brush. They are always there, these sounds, fleeting, passing by and then fading. We should prepare ourselves for receiving them, make a space for them and then wait. Listen… they will come.

Note

1. A suggested starting point by way of introduction might be Baron-Cohen, S., and Harrison, J. E., *Synaesthesia: Classic and Contemporary Readings.* Oxford: Blackwell, 1996.

Bibliography

Abse, Dannie. *A Poet in the Family.* London: Hutchinson, 1974.

Argyle, Michael. *The Psychology of Interpersonal Behaviour.* London: Penguin, 1994.

Aristotle. (Trans. Ross, G. R. T). *De Sensu and De Memoria.* Cambridge: Cambridge University Press, 1907.

————. (Trans. Lawson-Tancred, Hugh). *De Anima (On the Soul).* London: Penguin Books, 1986.

Arnheim, Rudolf. *Radio.* London: Faber and Faber, 1936.

Atkinson, Niall. 'The Republic of Sound: Listening to Florence at the Threshold of the Renaissance,' *I Tatti Studies in the Italian Renaissance,* Vol. 16. No. 1–2, pp. 57–84. Chicago: The University of Chicago Press on Behalf of Villa I Tatti. The Harvard Center for Italian Renaissance Studies, 2013. Source: https://www.jstor.org/stable/10.1086/673411,

Atkinson, Peter. 'The Sons and Heirs of Something Paricular: The Smiths' Manchester Aesthetic, 1982–1987,' in Franklin, Ieuan, Chignell, Hugh, and Skoog, Kristin (Eds.), *Regional Aesthetics: Mapping UK Media Cultures,* 71–89. Basingstoke. Palgrave Macmillan, 2015.

Attali, Jacques. *Noise: The Political Economy of Music.* Minneapolis: University of Minnesota Press, 2017.

Bachelard, Gaston. *The Poetics of Reverie*. Boston, MA: Beacon Press, 1971.
———. *The Poetics of Space*. Boston: Beacon Press, 1994.
Bajac, Quentin. *Being Modern: MoMA in Paris*. Paris and New York: Fondation Louis Vuitton and The Museum of Modern Art, 2017.
Baldwin, Stanley. *On England, and Other Addresses*. London: Philip Allen, 1926.
Barthes, Roland. *The Grain of the Voice*. London: Jonathan Cape, 1985.
———. *Camera Lucida*. London: Vintage, 2000.
Berger, John. *Ways of Seeing*. London: Penguin, 1972.
———. *About Looking*. London: Bloomsbury, 1980.
Bhaskar, Roy. *A Realist Theory of Science*. London: Verso, 2008.
Biewen, John, and Dilworth, Alexa. (Eds.). *Reality Radio: Telling True Stories in Sound*. Chapel Hill, NC: Duke University and University of Carolina Press, 2010.
Bishop, Elizabeth. *The Collected Prose*. New York: Farrar, Straus and Giroux, 1984.
Boston, Lucy M. *Memory in a House*. London: The Bodley Head, 1973.
———. *The Children of Green Knowe*. London: Faber and Faber, 2013.
———. *The River at Green Knowe*. London: Faber and Faber, 2013.
Bruller, Jean, alias 'Vercors' (Trans.) Connelly, Cyril (Ed.) Brown, James W. and Stokes, Lawrence D. *The Silence of the Sea*. London: Bloomsbury Academic, 2015.
Bull, Michael, and Black, Les. *The Auditory Culture Reader*. Oxford: Berg, 2003.
Burke, Edmund. *A Philosophical Enquiry into the Sublime and Beautiful*. Oxford: Oxford University Press, 2015.
Burrows, A. R. *The Story of Broadcasting*. London: Cassell, 1924.
Cantril, Hadley. *The Invasion from Mars: A Study in the Psychology of Panic*. Princeton: The Princeton University Press, 1952.
Carr, Samuel. (Ed.). *Ode to the Countryside*. London: National Trust Books, 2010.
Cavarero, Adriana. *For More Than One Voice: Toward a Philosophy of Vocal Expression*. Stanford, CA: Stanford University Press, 2005.
Chambers, Deborah. *Social Media and Personal Relationships: Online Intimacies and Networked Friendship*. Basingstoke: Palgrave Macmillan, 2013.
Chessa, Luciano. *Luigi Russolo, Futurist: Noise, Visual Arts and the Occult*. Berkeley: University of California Press, 2012.
Cluadel, Paul. *The Eye Listens*. Port Washington: Kennikat Press, 1969.
Coleridge, Samuel Taylor. *Coleridge: Poetry and Prose*. Oxford: Clarendon Press, 1954.
Connelly, Charlie. *Last Train to Hilversum: A Journey in Search of the Magic of Radio*. London: Bloomsbury, 2019.

Corbin, Alain. *Village Bells: Sound and Meaning in the Nineteenth-Century French Countryside.* Cambridge: Cambridge University Press, 1999.
———. *A History of Silence.* Cambridge: Polity Press, 2018.
Cox, Trevor. *Now You're Talking: The Story of Human Conversation from the Neanderthals to Artificial Intelligence.* London: The Bodley Head, 2018.
Crabbe, George. *The Poetical Works of George Crabbe.* Edinburgh: Gall and Inglis, 1854.
Crawford, Kate. 'Following You: Disciplines of Listening in Social Media,' in Sterne, Jonathan (Ed.), *The Sound Studies Reader,* 79–90. Abingdon: Routledge, 2012.
Crossley-Holland, Kevin. *Waterslain and Other Poems.* London: Hutchinson, 1986.
David, Francis. *The History of the Blues: The Roots, the Music, the People.* Boston, MA: Da Capo Press, 2003.
Deacon, George. *John Clare and the Folk Tradition.* London: Sinclair Browne, 1983.
De la Mare, Walter. *Collected Poems.* London: Faber and Faber, 1986.
Dickens, Charles. *Bleak House.* London: Chapman and Hall, 1903.
———. (Ed. Horsman, Alan). *Dombey and Son.* Oxford: Clarendon Press, 1974.
———. (Ed. Cardwell, Margaret). *The Mystery of Edwin Drood.* Oxford: Clarendon Press, 1972.
———. *David Copperfield.* London: Penguin, 1985.
———. *Hard Times.* London: Penguin, 2003.
Dickinson, Emily. *The Complete Poems of Emily Dickinson.* London: Faber and Faber, 1977.
———. (Ed. Fragos, Emily). *Emily Dickinson: Letters.* New York: Everyman's Library Pocket Poets, 2011.
Dillard, Annie. *Pilgrim at Tinker Creek.* London: HarperCollins, 2011.
Douglas, Keith. (Ed. Graham, Desmond). *Keith Douglas: The Letters.* Manchester: Carcanet, 2000.
Drabble, Margaret. *A Writer's Britain.* London: Thames and Hudson, 2009.
Esslin, Martin. *Mediation: Essays on Brecht, Beckett and the Media.* London: Eyre Methuen, 1980.
Evans, George Ewart. *Spoken History.* London: Faber and Faber, 1987.
Evelyn, John. (Ed. Francs, Philip). *John Evelyn's Diary.* London: The Folio Society, 1963.
Ferneyhough, Charles. *The Voices Within: The History and Science of How We Talk to Ourselves.* London: Profile Books, 2016.
Fitzgerald, Penelope. *The Blue Flower.* London: Flamingo, 2003.

Fontana, David. 'Altered States Through Meditation and Dreams,' in Lorimer, David (Ed.), *Thinking Beyond the Brain: A Wider Science of Consciousness*. Edinburgh: Floris Books, 2001.

Friedrich, Otto. *Glenn Gould: A Life and Variations*. London: Lime Tree, 1990.

Frost, Robert. (Ed. Thompson, Lawrance). *Selected Letters of Robert Frost*. New York: Holt, Rinehart and Winston, 1964.

Gann, Kyle. *No Such Thing as Silence: John Cage's 4'33"*. New York: Yale University, 2010.

Gilfillan, Daniel. *Pieces of Sound: German Experimental Radio*. Minneapolis: University of Minnesota Press, 2009.

Godwin, Joscelyn. *Athanasius Kircher's Theatre of the World: His Life, Work, and the Search for Universal Knowledge*. Rochester, VT: Inner Traditions, 2015.

Gormley, Antony, Richardson, Clare, and Winterson, Jeanette. *Land*. Maidenhead: The Landmark Trust, 2016.

Goswami, Amit. *The Self-Aware Universe: How Consciousness Creates the Material World*. New York: Jeremy P. Tarcher/Putnam, 1995.

Gracq, Julien. (Trans. Richard, Howard). *The Opposing Shore (Le Rivage des Syrtes)*. London: Harvill, 1993.

Greenfield, Susan. *The Human Brain: A Guided Tour*. London: Weidenfeld and Nicolson, 1997.

Harris, Jose. *Private Lives, Public Spirit: Britain, 1870–1914*. London: Penguin, 1993.

Harrison, John. *Synaesthesia: The Strangest Thing*. Oxford: Oxford University Press, 2007.

Haslam, Dave. *Manchester England: The Story of the Pop Cult City*. London: Fourth Estate, 1999.

Haydn, Eleanor. *Travels Around Our Village*. London: Archibald Constable, 1905.

Hazan, Eric. *The Invention of Paris: A History in Footsteps*. London: Verso, 2010.

Hazlitt, William. (Ed. Vallance, Rosalind, and Hampden, John). *Essays*. London: Folio Society, 1964.

Heaney, Seamus. *New Selected Poems 1966–1987*. London: Faber and Faber, 1990.

———. *Finders Keepers: Selected Prose 1971–2001*. London: Faber and Faber, 2002.

Hempton, Gordon. *One Square Inch of Silence: One Man's Quest to Preserve Quiet*. New York: Free Press, 2009.

Hoban, Russell. *The Moment Under the Moment*. London: Picador, 1993.

Howard, Deborah, and Moretti, Laura. *Sound and Space in Renaissance Venice*. New Haven: Yale University Press, 2009.

Hughes, Richard. *Plays*. London: Chatto and Windus, 1928.

Hull, John M. *Touching the Rock*. London: SPC, 2016.

Ingelow, Jean. *Poems*. Boston: Roberts Brothers, 1863.

Jefferies, Richard. *Field and Hedgerow*. London: Lutterworth Press, 1948.

———. *Round About a Great Estate*. Bradford on Avon: Ex Libris Press, 1987.

Johnson, William. (Ed.). *Focus on the Science Fiction Film*. Englewood Cliffs, NJ: Prentice Hall, 1972.

Jolly, W. P. *Sir Oliver Lodge: Psychical Researcher and Scientist*. London: Constable, 1974.

Jones, David. *The Anathema*. London: Faber, 1972.

Joyce, James. *Dubliners*. Ware: Wordsworth Editions, 1993.

Kafka, Franz. *The Blue Octavo Notebooks*. Cambridge: Exact Change, 1993.

Kagge, Erling. *Silence in the Age of Noise*. London: Penguin, 2017.

Kandinsky, Wassily. *Complete Writings on Art*. New York: Da Capo Press, 1994.

Keats, John. *Poetical Works*. London: Oxford University Press, 1967.

Le Guin, Ursula. *Steering the Craft: A 21st Century Guide to Sailing the Sea of the Story*. New York: Mariner Books, 2015.

Leighton, Angela. *Hearing Things: The Work of Sound in Literature*. Cambridge, MA: The Belknap Press of Harvard University Press, 2018.

Lester, Julius. *To Be a Slave*. New York: Puffin Books, 2005.

Levitin, Daniel. *This Is Your Brain on Music: Understanding a Human Obsession*. London: Atlantic, 2006.

Lewis, C. A. *Broadcasting from Within*. London: George Newnes, 1924.

Locke, John. *An Essay on Human Understanding*. Oxford: Oxford University Press, 2008.

Lodge, Oliver. *My Philosopy: Representing My Views on the Many Functions of the Ether of Space*. London: Ernest Benn, 1933.

Macfarlane, Robert. *Landmarks*. London: Penguin Books, 2016.

MacPherson, James. *Fingal and Other Poems of Ossian*. Victoria: Leopold Classic Library, 2018.

Mansell, James G. *The Age of Noise in Britain*. Chicago: University of Illinois Press, 2017.

Massey, Irving J. 'The Musical Dream Revisited: Music and Language in Dreams,' *Psychology of Aesthetics, Creativity, and the Arts*, Vol. S(1), pp. 42–50, 2006.

Mazierska, Ewa. (Ed.). *Northern Sounds: Popular Music, Culture and Place in England's North*. Sheffield: Equinox, 2018.

McWhinnie, Donald. *The Art of Radio*. London: Faber and Faber, 1959.

Merleau-Ponty, Maurice. *Phenomenology of Perception*. London: Routledge, 2002.

Millier, Brett C. *Elizabeth Bishop: Life and the Memory of It.* Berkeley: University of California Press, 1993.

Motion, Andrew. *Keats.* London: Faber and Faber, 1998.

Muers, Rachel. *Keeping God's Silence: Towards a Theological Ethics of Communication.* Oxford: Blackwell, 2004.

Nancy, Jean-Luc. (Trans. Mandell, Charlotte). *Listening.* New York: Fordham University Press, 2007.

Niebur, Louis. *Special Sound: The Creation and Legacy of the BBC Electronic Workshop.* Oxford: Oxford University Press, 2010.

Oakley, Giles. *The Devil's Music: A History of the Blues.* Boston, MA: Da Capo Press, 1997.

Ong, Walter. *The Presence of the Word: Some Prolegomena for Cultural and Religious History.* Minneapolis: University of Minnesota Press, 1967.

Payzant, Geoffrey. *Glenn Gould: Music and Mind.* Toronto: Van Nostrand Reinhold, 1978.

Pepys, Samuel. *Diary.* London: Marshall Cavendish, 1988.

Picker, John M. *Victorian Soundscapes.* Oxford: Oxford University Press, 2003.

Price, Barnard. *Creative Landscapes of the British Isles.* London: Ebury Press, 1983.

Prochnik, George. *In Pursuit of Silence: Listening for Meaning in a World of Noise.* New York: Anchor Books, 2010.

Proust, Marcel. (Trans. Scott Moncrieff, C. K. and Kilmartin, Terence). *In Pursuit of Lost Time, Vol. 1: Swann's Way.* London: Vintage, 2002.

Reith, J. C. W. *Broadcast over Britain.* London: Hodder and Stoughton, 1924.

Renner, Rolf G. *Hopper.* Cologne: Taschen, 2011.

Robinson, Tim. *Connemara.* London: Penguin Books, 2007.

Rodenburg, Patsy. *The Right to Speak: Working with the Voice.* London: Methuen Drama, 2005.

Rollins, Hyder E. (Ed.). *The Letters of John Keats, Volume II: 1814–1821.* Cambridge: Cambridge University Press, 1958.

Rowlands, Peter. *Oliver Lodge and the Liverpool Physical Society.* Liverpool: Liverpool University Press, 1990.

Rushing, Wanda. 'We're Going to Graceland: Globalisation and the Reimagining of Memphis,' in Lashua, Brett, Spracklen, Karl, and Wag, Stephen (Eds.), *Sounds and the City: Popular Music, Place and Globalisation.* Basingstoke: Palgrave Macmillan, 2014.

Russell, Tony. *Blacks, Whites and Blues.* London: Studio Vista, 1970.

Russolo, Luigi, et al. *The Art of Noise: Destruction of Music by Machines.* London: Sun Vision Press, 2012.

Sacks, Oliver. *Musicophilia: Tales of Music and the Brain.* London: Picador, 2012.

Schnupp, Jan, Nelken, Israel, and King, Andrew. *Auditory Neuroscience: Making Sense of Sound.* Cambridge, MA: The MIT Press, 2011.

Schwarz, A. Brad. *Broadcast Hysteria: Orson Welles's "War of the Worlds" and the Art of Fake News.* New York: Hill and Wang, 2015.

Sconce, Jeffrey. *Haunted Media: Electronic Presence from Telegraphy to Television.* Durham: Duke University Press, 2007.

Seargeant, Philip, and Tagg, Caroline. (Eds.). *The Language of Social Media: Identity and Community on the Internet.* Basingstoke: Palgrave Macmillan, 2014.

Seibert, Brian. *What the Eye Hears: A History of Tap Dancing.* New York: Farrar, Straus and Giroux, 2015.

Shepherd, Nan. *The Living Mountain in the Grampian Quartet.* Edinburgh: Canongate, 1996.

Shute, John. *The First and Chief Groundes of Architecture (Facsimile, 1563 Folio).* London: Country Life, 1912.

Sieveking, Lance. *The Stuff of Radio.* London: Cassell, 1934.

Small, Christopher. *Music of the Common Tongue: Survival and Celebration in Afro-American Music.* London: Calder, 1987.

Smith, Bruce R. *The Acoustic World of Early Modern England: Attending to the O Factor.* Chicago: University of Chicago Press, 1999.

Smith, Mark R. *Hearing History: A Reader.* Athens, Georgia: University of Georgia Press, 2004.

Smith, Patti. *Devotion.* New Haven: Yale University Press, 2017.

Sonnenschein, David. *Sound Design: The Expressive Power of Music, Voice, and Sound Effects in Cinema.* Studio City, CA: Michael Wiese Productions, 2001.

Spinelli, Martin, and Dann, Lance. *Podcasting: The Audio Media Revolution.* London: Bloomsbury, 2019.

Stanev, Hristomir A. *Ben Jonson's Eloquent Nonsense: The Noisy Ordeals of Heard Meanings on the Jacobean Stage (1609–14).*http://dx.doi.org/10.12745/et.17.2.1211.

Steinbeck, John. *East of Eden.* London: Penguin, 2000.

Stenger, Susan. *Sound Strata of Coastal Northumberland.* Newcastle: AV Festival, 2014.

Stewart, Garrett. *Reading Voices: Literature and the Phonotext.* Berkeley: University of California Press, 1990.

Stocker, Michael. *Hear Where We Are: Sound, Ecology and Sense of Place.* New York: Springer, 2013.

Street, Seán. *Radio Waves: Poems Celebrating the Wireless.* London: Enitharmon Press, 2004.

————. *The Memory of Sound: Preserving the Sonic Imagination.* Abingdon: Routledge, 2015.

Takemitsu, Toru. *Confronting Silence: Selected Writings.* Berkeley: Falling Leaf Press, 1995.

Tennyson, Alfred. *The Works of Alfred, Lord Tennyson.* London: Macmillan, 1926.

Thomas, Edward. *The Icknield Way.* London: Constable, 1913.

————. *Oxford.* London: A & C Black, 1913.

————. *In Pursuit of Spring.* Toller Fratrum: Little Toller Books, 2016.

Thompson, Paul. *The Voice of the Past: Oral History.* Oxford: Oxford University Press, 1988.

Thoreau, Henry David. *The Journal, 1837–1861.* New York: New York Review Books, 2009.

————. *Walden.* London: Penguin, 2016.

Tomalin, Norman. *Daventry Calling the World.* Whitby: Caedmon of Whitby Publishers, 1998.

Toop, David. *Sonic Boom: The Art of Sound.* London: Hayward Gallery, 2000.

————.*Sinister Resonance: The Mediumship of the Listener.* New York: Continuum, 2010.

Tuer, Andrew W. *Old London Street Cries.* London: The Scolar Press, 1978.

'Vercors'. (Ed. Brown, James W. and Stokes, Lawrence D). *The Silence of the Sea (Le Silence de la Mer).* London: Bloomsbury Academic, 2015.

Vitruvius. (Trans. Morgan, M. H.). *The Ten Books on Architecture.* New York: Dover Publications, 1960.

Webb, Jeff A. *The Voice of Newfoundland: A Social History of the Broadcasting Corporation of Newfoundland, 1939–1949.* Toronto: University of Toronto Press, 2008.

Weinstein, Deena. 'Birmingham's Post-industrial Metal,' in Lashua, Brett, Spracklen, Karl, and Wagg, Stephen (Eds.), *Sounds and the City: Popular Music, Place and Globalization,* 38–54. Basingstoke: Palgrave Macmillan, 2014.

Weiss, Elisabeth, and Belton, John. *Film Sound: Theory and Practice.* New York: Columbia University Press, 1985.

Westerkamp, Hildegard. 'The Soundscape on Radio,' in Augaitis, Daina and Lander, Dan (Eds.), *Radio Rethink: Art, Sound and Transmission.* Banff: Walter Phillips Gallery, 1994.

White, Gilbert. *The Natural History of Selborne, with a Preface by Richard Jefferies.* London: Walter Scott, undated.

Whyte, William. *Unlocking the Church: The Lost Secrets of Victorian Sacred Space.* Oxford: Oxford University Press, 2017.

Winterson, Jeanette. (With Gormley, Antony and Richardson, Clare). *Land.* Maidenhead: The Landmark Trust, 2016.

Wordsworth, William. *The Poetical Works of William Wordsworth.* London: Henry Frowde and Oxford University Press, 1909.

Yeats, W. B. *Essays and Introductions.* New York: Macmillan, 1961.

Young, William H., and Young, Nancy K. *Music of the Great Depression.* Westport, CT: Greenwood Press, 2005.

Index

The manufacturer's authorised representative in the EU is Springer
Nature Customer Service Centre GmbH, Europaplatz 3, 69115 Heidelberg,
Germany. If you have any concerns regarding our products, please
contact ProductSafety@springernature.com

Printed and bound by CPI Group (UK) Ltd, Croydon, CR0 4YY
29/04/2026
02099478-0008